Biological and Medical Physics, Biomedical Engineering

This series is intended to be comprehensive, covering a broad range of topics important to the study of the physical, chemical and biological sciences. Its goal is to provide scientists and engineers with monographs- both, authored and contributed volumes to address the growing need for information. The fields of biological and medical physics and biomedical engineering are broad, multidisciplinary and dynamic. They lie at the crossroads of frontier research in physics, biology, chemistry, and medicine.

Books in the series emphasize established and emergent areas of science including molecular, membrane, and mathematical biophysics; photosynthetic energy harvesting and conversion; information processing; physical principles of genetics; sensory communications; automata networks, neural networks, and cellular automata. Equally important is coverage of applied aspects of biological and medical physics and biomedical engineering such as molecular electronic components and devices, biosensors, medicine, imaging, physical principles of renewable energy production, advanced prostheses, and environmental control and engineering.

Alexei V. Chvetsov

Optimization Problems in Radiotherapy for Hypoxic Tumors

Theory and Applications

Springer

Alexei V. Chvetsov
Department of Radiation Oncology
University of Washington
Seattle, WA, USA

ISSN 1618-7210 ISSN 2197-5647 (electronic)
Biological and Medical Physics, Biomedical Engineering
ISBN 978-981-96-7000-0 ISBN 978-981-96-7001-7 (eBook)
https://doi.org/10.1007/978-981-96-7001-7

This Springer imprint is published by the registered company Springer Nature Singapore Pte Ltd.
The registered company address is: 152 Beach Road, #21-01/04 Gateway East, Singapore 189721, Singapore

Preface

Radiation therapy is one of the main modalities to cancer treatment that is rapidly developing using the advancement in medical imaging, high-capacity computers, and radiation biology. The rapid development of cancer radiation therapy stimulated active search for effective computational methods for treatment outcomes analysis, biological modeling, and dose optimization. This topic was addressed in several review articles and monographs. Nevertheless, the mathematical aspects of treatment outcomes analysis and dose optimization in radiotherapy for hypoxic tumors did not receive a systematic consideration. We note that hypoxia is present in most tumors encountered in humans.

Tumor hypoxia is characterized by regions with low partial oxygen pressure (pO_2) and is a major factor of cancer resistance to radiotherapy with sparsely ionizing radiation such as high-energy photons and electrons. The presence or absence of molecular oxygen strongly influences the biological effect of sparsely ionizing radiation because oxygen makes DNA damage permanent and irreparable. There are no established computational methods in cancer radiotherapy to describe hypoxia in treatment outcomes analysis or how to overcome hypoxia radio-resistance. There are several challenges for considering hypoxia in radiotherapy analysis and optimization, and one of them is the complex patient-specific 3D distributions of pO_2 within the tumors leading to patient-specific variation of cell killing effect. Therefore, the variation of biological cell killing effect between patients is not considered, and the standard of care is to prescribe the same radiation dose for all patients independently of tumor hypoxia.

The authors systematize the information on mathematical problems in radiation therapy for hypoxic tumors and analyze important aspects and possible trends in the field. The book contains new approaches to the concepts of tumor control probability, equivalent uniform dose and radiotherapy dose optimization for hypoxic tumors developed by the author. Significant attention in this book is paid to comparison of models with measured and clinical data; therefore, the reduction of model parameters to overcome overfitting (model parsimony) was followed as much as possible. The impact of hypofractionation as one of the main trends in development of radiotherapy was also included in the analysis. The authors did not try to present a comprehensive review of research results achieved in computational and biophysical methods for radiotherapy analysis and optimization. The authors only formulated their own point of view on the topics of their interest and attention in the biophysical modeling for hypoxic tumors. The authors assumed that many other aspects of biophysical modelling in radiation therapy have been considered in the monographs of other authors, and these topics may include relative biological effectiveness of high-LET (linear energy transfer) radiation, normal tissue complication probabilities, and application of machine learning in treatment outcomes analysis.

The goal of this book is to draw the attention of researchers with an interest in computational mathematics to the variety of mathematical problems encountered in radiotherapy for hypoxic tumors. It may help clinical and theoretical medical physicists in the analysis of radiotherapy outcomes and the development of new radiotherapy approaches for hypoxic tumors. The authors tried to present a comprehensive introduction in radiotherapy for hypoxic tumors for a reader, but the emphasis was put on mathematical content of the problems.

The content of this book is the result of author's research activities in clinical medical physics in collaboration with physicists, physicians, and researchers in the Department of Radiation Oncology and Molecular Imaging Center in the Department of Radiology at the University of Washington, USA. The author would like to express the acknowledgements to collaborators from these and other institutions who contributed significantly to this monograph and helped the author: Daniella Calvetti, Ph.D., Mark Muzi, M.S., Lei Dong, Ph.D., Jeffery L. Schwartz, Ph.D., Robert D. Stewart, Ph.D., Leonid G. Hanin, Ph.D., Joseph Rajendran, M.D., Ramesh Rengan, M.D., Ph.D., Jing Zeng, M.D., Nina A. Mayr, M.D., Simon S. Lo, M.D., George A. Sandison, Ph.D., Jatinder J. Palta, Ph.D., Stephen R. Bowen, Ph.D., Andrei Pugachev, Ph.D.

The chapters of the book can be summarized as follows:

Chapter 1 presents a general introduction to biomathematical problems that may be needed in treatment outcomes analysis and optimization in radiotherapy for hypoxic tumors with the emphasis on hypoxia imaging, cell survival models and mechanistic models of tumor control probability.

Chapter 2 describes the multi-level cellular radiotherapy response models that are needed for the analysis and optimization of tumor response to radiotherapy, parameter estimation and development of new radiotherapy techniques. Special attention is paid to model validation by comparison with the results of clinical trials. The radiotherapy response models include a sum of several exponentials describing cell proliferation, cell killing, and cell clearance; therefore, the problem of data fitting is ill-posed and may produce large nonphysical fluctuations in reconstructed radiobiological parameters. We describe a regularization technique to overcome the problem of ill-posedness and discuss parameters' estimation in radiotherapy for head-and-neck cancer.

Chapter 3 describes a derivation of equations for tumor control probability (TCP) equation for tumors with hypoxic regions. It is shown that the dose-volume effects in the TCP are dependent on tumor radiosensitivity and defined primarily by the hypoxic regions. A comparison with clinical data is presented for both conventional and hypofractionated radiotherapies.

In Chap. 4 we generalize the concept of equivalent uniform dose (EUD) to hypoxic tumors and introduce a new concept of equivalent uniform aerobic dose (EUAD) that is needed for analysis and optimization of treatment outcomes in radiotherapy for heterogeneous hypoxic tumors with 3D variable radiosensitivity. The concept of equivalent uniform dose (EUD) has been proven instrumental for the analysis of treatment outcomes and biological optimization of radiotherapy. However, the concept of EUD does not consider hypoxic tumors because it was developed for oxygenated tumors. The new concept of EUAD assumes that for any non-uniform distributions of the Oxygen Enhancement Ratio (OER) and radiation dose within a tumor, there is an equivalent uniform distribution of radiation dose under hypothetical aerobic conditions with OER=1 that produces equal survival of clonogenic cells. The new concept of EUAD assumes that two dose distributions in hypoxic tumors are equivalent if they produce equal biological effect under aerobic conditions.

The goal of Chap. 5 is to determine the maximum gain in biological effectiveness that can be achieved using hypoxia targeted dose escalation protocols compared to the standard (for example uniform) dose distributions. To quantitatively describe the biological effect of non-uniform dose distributions, we define the dose non-uniformity effectiveness (DNE) as ratio of the EUAD for non-uniform dose distribution and uniform dose distributions assuming that they are computed for the same integral physical dose. We compute DNE for different values of the OER, hypoxic and tumor volumes and determine where the non-uniform dose distributions may be feasible.

Chapter 6 describes the advanced dose optimization methods that can be used to compute the optimal OER-based dose distributions using direct and in-direct optimization of the EUAD. The EUAD is needed to derive non-uniform dose distributions in the hypoxic tumors that are expected to improve the therapeutic ratio of radiation therapy. The optimized beam intensity functions which produce the optimal dose distributions are obtained using algorithms of inverse treatment planning that is an ill-posed problem. To stabilize the numerical solutions, special regularization techniques have been developed. We propose to customize these techniques for inverse treatment planning algorithms to improve quality of numerical solutions and, in turn, the modality of dose delivery to the cancer tumor. We apply to the inverse treatment planning the L-curve method which was developed to determine the regularization parameter in the discrete ill-posed problems.

Seattle, USA Alexei V. Chvetsov

Competing Interests The author has no competing interests to declare that are relevant to the content of this manuscript.

Contents

Chapter 1
Introduction

1.1 Biomarkers for Individualized Dose Prescriptions

We consider biomathematical modeling and optimization of cancer radiotherapy assuming that the tumor response to radiotherapy is defined by the following three broad radiobiological categories: [1]: (1) intrinsic cellular radiosensitivity, (2) cellular proliferative potential, and (3) tumor oxygen status. The intrinsic cellular radiosensitivity, cellular proliferative potential, and tumor oxygen status vary between patients and markedly contribute to inter-patient variation of the tumor control probability (TCP) and treatment outcomes. The prognostic value of biomarkers associated with these radiobiological determinants of radiotherapy response has been demonstrated in many research studies [1–9].

Despite inter-patient variation in radiotherapy response the same population-based radiation dose prescriptions are currently used for radiotherapy treatments. The individualized optimization of biological effects has the potential to reduce morbidity associated with ineffective treatments and to improve radiotherapy outcomes for cancers with low survival rates. For example, the survival rates are low in radiotherapy for glioblastoma, the most common and most aggressive type of primary malignant brain tumors [8]. The median overall survival in patients with glioblastoma is less than 15 months despite treatments with surgery, radiotherapy and chemotherapy. Another example of low survival rates is radiotherapy for stage III non-small cell lung cancer [9].

Clinical implementation of individualized treatments depends on effective biomarkers to predict and evaluate tumor response to radiotherapy treatments; therefore, development of biomarkers is an important and ongoing effort in oncology. For predicting radiation treatment responses, early biomarker studies focused on ex vivo determination of inherent radiation sensitivity, measurement of tumor proliferative potential, and electrode-based analysis of hypoxia. While these assays showed some predictive value, they were labor intensive and invasive. Later developments focused on noninvasive functional imaging and gene expression biomarkers.

A. V. Chvetsov, *Optimization Problems in Radiotherapy for Hypoxic Tumors*, Biological and Medical Physics, Biomedical Engineering,
https://doi.org/10.1007/978-981-96-7001-7_1

(a) **Intrinsics Cellular Radiosensitivity**

For example, intrinsic tumor radiosensitivity characterized by cell surviving fractions is a key parameter in the evaluation of tumor response to radiotherapy and can be evaluated using predictive assays [1–3]. The extensive data on patient-specific cell survival fraction S for the initial part of dose–response curve have been measured in vitro using predictive assays, and their prognostic values for the treatment outcome have been demonstrated. For example, clonogenic cellular assays in vitro were used to generate a few published data sets for cell surviving fraction S_2 after a 2 Gy dose for different types of tumors. The surviving fraction S_2 has been reported to be a predictor of treatment outcomes; for example, several research groups have found that survival of patients with S_2 above the median value was significantly poorer than those below the median value [1–3]. However, practical use of predictive assays of tumor radiosensitivity is hampered by a high failure rate, labor intensiveness, and inability to evaluate multimodality treatments that include chemotherapy and radiotherapy. Application of these biomarkers for treatment modification and optimization would require higher level of uncertainties and reliability because the cell population of shrinking irradiated tumors is not expected to behave in a manner similar to tumor cell cultures obtained using tumor biopsy. Therefore, there are ongoing studies to develop alternatives to in vitro predictive assays.

More recent work in identifying and assessing gene expression biomarkers of tumor response is very promising [4, 5]. However, like the other ex vivo assays, gene expression biomarkers do not provide information on regional variations in tumor treatment response that are associated with, for example, hypoxia. Multimodality treatments might be a complication that reduces the power of these gene-based assays.

(a) **Cellular Proliferation**

Cellular proliferative potential can be measured in vitro using clonogenic assays like intrinsic cellular radiosensitivity. Additionally, PET biomarkers such as 3'-deoxy-3'-[18F] fluorothymidine (FLT) have been developed for the in vivo evaluation of proliferation and its spatial distribution [6]. We note that the tumor cells proliferation is significant in the oxygenated regions and less significant in the hypoxic regions that are considered a major factor in tumor radioresistance. Cellular proliferation potential in tumors during radiotherapy may have a comparatively smaller effect on the TCP compared with the effects of variation in radiosensitivity.

(b) **3D Hypoxia Measurements and Dose Optimization**

Treatment outcomes analysis shows that the local failures in many tumors may be due to increased radioresistance caused by hypoxic regions. Tumor hypoxia has been initially measured using invasive electrode and poorer outcomes for hypoxic tumors have been demonstrated [7]. While these assays showed some predictive value, they were labor intensive and invasive. Significant progress in this area was achieved using functional imaging based on the positron emission tomography (PET) and magnetic resonance imaging (MRI). Noninvasive functional imaging has the potential to identify local variations in tumor environment and the development of resistance in tumors during multimodality treatments. The noninvasive PET and MR imaging can measure the 3D distributions of hypoxia in tumors that can be used in outcomes analysis and patient-specific biological optimization of radiotherapy treatment [10]. Using PET and MR imaging with different biomarkers, it has been shown for different cancers including glioblastoma, cervical cancer and head and neck cancer that increased levels of chronic tumor hypoxia are associated with poor treatment outcomes [8, 11–13].

PET and MRI hypoxia imaging present a most promising pathway for biological radiotherapy optimization because it provides 3D information about hypoxia distribution in noninvasive way before the treatment begins [14–16]. This type of hypoxia imaging has the potential to identify local variations in tumor environment and the development of resistance in tumors during multimodality treatments. The clinical implementation of radiotherapy treatment machines that integrate medical linear accelerators with PET or MRI could increase the application of hypoxia imaging in radiation oncology and potentially improve treatment outcomes [17].

We note that the absolute value of tumor-specific radiosensitivity defined by cell survival cannot be measured using available modalities of anatomical or functional imaging; therefore, a population value should be assumed for the intrinsic radiosensitivity. However, it should not be a problem in biological optimization because the major patient-specific factor in tumor radio-resistance is defined by the oxygen effect.

It is also useful to mention that the widely used biomarker ^{18}F-Flurodeoxyglucose (FDG) usually associated with the glucose metabolism shows a positive correlation with hypoxia [18, 19]. This correlation was demonstrated in the comparison with hypoxia measurements imaging and autoradiography. It is an indication of the fact that hypoxia results in accelerated glycolysis despite the cellular metabolism slowing down in chronic hypoxia. In general, the FDG uptake may depend on both hypoxia and cellular metabolism and may be very tumor type specific.

1.2 Oxygen Effect and Tumor Hypoxia

1.2.1 Models for Oxygen Enhancement Ratio

The oxygen effect is an important factor in radiotherapy because tumors may contain a significant fraction of clonogenic cells that exist at low pO_2 to provide significant hypoxic radioprotection. About two-thirds of the biologic damage by means of the sparsely ionizing radiation such as X-rays, protons, and electrons is produced by indirect action of free radicals [1]. The damage in DNA produced by free radicals can be repaired under hypoxia, but presence of molecular oxygen makes DNA damage permanent and irreparable.

The biological effect of different levels of pO_2 on cell survival is quantitatively described by the oxygen enhancement ratio (OER). Originally, the OER was defined as a ratio of the dose under hypoxic or anoxic conditions d_{hyp} to the dose d under variable pO_2 needed to achieve the same biological effect [1, 20, 21]:

$$OER_{hyp} = \frac{d_{hyp}\left(pO_{2,\min}\right)}{d\left(pO_2\right)}, \tag{1.1}$$

where $pO_{2,min}$ is the minimum pO_2 corresponding to anoxic conditions. It was shown that experimental measurements of OER_{hyp} may be described by the following function [20, 21]:

$$OER_{hyp}\left(pO_{2,}\right) = \frac{OER_{\max} \cdot pO_{2,} + K}{pO_2 + K} \tag{1.2}$$

where $OER_{max} = d_{hyp}/d_{ox}$. is the maximum value achieved under aerobic conditions and K is pO_2 \at which half-maximum sensitization is achieved. The maximum arterial pressure of oxygen is typically 100 mmHg, and the necrosis threshold is below 1 mmHg. The validity of Eq. 1.2 was tested in several independent measurements [21].

Equations 1.1 and 1.2 assume the anoxic conditions for radiosensitivity as the initial point for computing the OER_{hyp}; therefore, the minimum value of the OER_{hyp} = 1 is achieved under anoxic conditions and the maximum value of the OER_{hyp} is achieved under well-oxygenated conditions. This formulation of the OER_{hyp} given by Eq. 1.1 is difficult to apply for computing the escalated doses in hypoxic tumor regions that produce equivalent biological effect observed in well-oxygenated regions. Therefore, a modified definition of the OER was proposed earlier where the initial point for computing the OER was considered at well-oxygenated conditions [22, 23]. This modified OER is defined as a ratio of the dose under variable pO_2 to the dose under well-oxygenated conditions needed to achieve equal biological effect:

$$OER = \frac{d\left(pO_2\right)}{d_{ox}\left(pO_{2,\max}\right)}, \tag{1.3}$$

where $pO_{2,max}$ is the maximum arterial pO_2. Using Eqs. 1.1 and 1.3 the modified OER can be obtained as a function of the OER_{hyp}:

$$OER = \frac{OER_{\max}}{OER_{hyp}\left(pO_2\right)}. \tag{1.4}$$

Substituting Eq. 1.2 in Eq. 1.4, we obtain the modified OER as a function of pO_2 [24]:

$$OER\left(pO_{2,}\right) = \frac{OER_{\max} \cdot pO_{2,} + OER_{\max} \cdot, K}{OER_{\max} \cdot pO_{2,} + K}. \tag{1.5}$$

A slightly different empirical equation was previously used for the modified OER instead of Eq. 1.5 [22, 23, 25]:

Fig. 1.1 Different models of OER as a function of pO_2

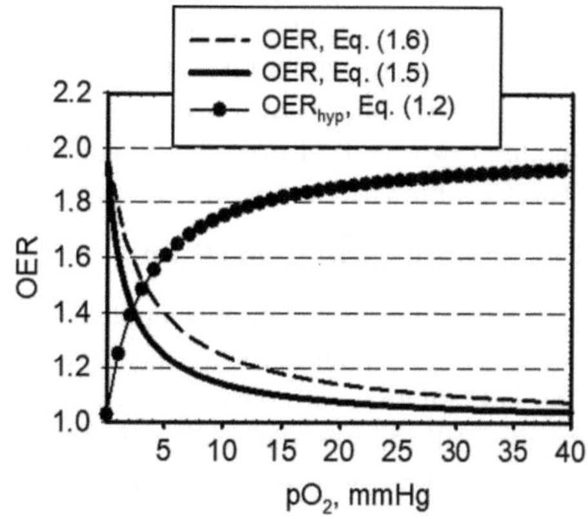

$$OER(pO_{2,}) = \frac{pO_{2,} + OER_{\max} \cdot, K}{pO_{2,} + K}. \tag{1.6}$$

A comparison of the OER as a function of pO_2 computed using Eqs. 1.2, 1.5 and 1.6 is shown in Fig. 1.1. It was assumed that $OER_{max} = 2.0$ and K = 3.28 mmHg in these calculations. Equations 1.5 and 1.6 for the OER have similar asymptotic approximations:

$$\lim_{pO_2 \to 0} OER(pO_{2,}) = OER_{\max}, \tag{1.7}$$

$$\lim_{pO_2 \to \infty} OER(pO_{2,}) = 1. \tag{1.8}$$

However, Eq. 1.6 overestimates Eq. 1.5 for equal values of the OER_{max} and K. The modified definition of OER assumed in Eqs. 1.3-1.6 is more convenient for practical applications because the maximum value of the OER is achieved under anoxic conditions and minimum value of OER = 1 is achieved under well-oxygenated conditions. The OER computed using this model can be considered as a dose escalation factor needed to achieve the same biological effect in hypoxic regions observed for well-oxygenated cells.

1.2.2 Effects of Radiation Dose and Ionization Density

(a) OER and Radiation Dose

The measured maximum value of OER under hypoxic conditions is between 2.5 and 3.5 for sparsely ionizing radiation and it is dependent on the dose of radiation. In vitro measurements demonstrate that the OER has a smaller value of about 2.5 at lower doses comparable to daily doses per fraction used in conventional radiation therapy and approaches 3.5 when the dose per fraction increases [1, 21]. This effect may stem from the variation of the OER with the phase of the cell cycle. It is believed that cells in G_1 phase have lower value of OER compared to S phase and they dominate the low region of the cell survival curve because they are more radiosensitive. Therefore, the OER is slightly smaller at low doses than at high doses.

(b) **OER and Ionization Density**

The density of ionization is another factor that affects the value of the OER. The oxygen effect is smaller for densely ionizing radiation such as alpha particles, carbon ions and radiation with intermediate density such as neutrons [1]. It has been shown that the OER reduces to much smaller values compared to sparsely ionizing radiation and is approximately 1.6 tor radiation with intermediate ionization density such as neutrons. For densely ionizing radiation such as low energy alpha particles, the OER approaches unity that means there is no oxygen effect. We note that the strength of oxygen effect is related to the magnitude of shoulder in the survival curve. The shoulder is reduced for neutrons and is not present for the low-energy alpha particle.

In theory and simulations presented in this book, we consider the dependence of radiotherapy response for hypoxic tumors on radiation dose; however, we review only sparsely ionizing radiation such as photons, electrons, and protons. Most radiotherapy treatments are currently delivered using sparsely ionizing radiation.

1.2.3 Chronic and Acute Tumor Hypoxia

Tumor hypoxia characterized by regions with low levels of pO_2 is a major factor of cancer resistance to radiotherapy with sparsely ionizing radiation. The hypoxic cells are of primary concern in radiotherapy because they are viable but sufficiently radioresistant. The increased radioresistance to radiotherapy is a primary concern associated with hypoxia, but it is useful to mention that there is clinical and laboratory evidence of the correlation between the low level of tumor pO_2 and increased risk of distant metastatic potential of solid tumor cells [26].

(a) **Chronic or Diffusion Limited Hypoxia**

One of the two main causes of hypoxia in human tumors is known as ‘chronic” or “diffusion-limited” hypoxia and is due to the lack of sufficient numbers of tumor blood vessels. The rapid proliferation of tumor blood cells results in a demand for oxygen that exceeds the capacity of the tumor vascular network. The pO_2 declines with distance from capillaries, partly because of the metabolism of intervening cells and partly due to diffusion. Calculations show that the “chronic” hypoxia indicates the existence of cells at all possible oxygen concentrations from normal values next to the vessels to necrotic cells at zero oxygen concentration at distances greater than 150 μm from the capillaries, with a layer of hypoxic cells between them [1].

(b) **Acute or Perfusion Limited Hypoxia**

The second main reason of hypoxia in tumors is the acute or 'perfusion-limited" hypoxia that is characterized by rapidly changing oxygen concentration in areas where the blood flow through the vessels is unstable. As a result, the oxygen concentration can vary between normal for well-perfused cells and anoxia for complete vessel blockage. There is good evidence that tumor blood vessels close and open in a random fashion so that different so that different regions of tumor become hypoxic intermittently [1, 10].

Acute or "perfusion" hypoxia is a dynamic process that is highly unpredictable and can occur at any time during radiation; therefore, inclusion of acute hypoxia in the analysis of treatment outcomes and radiotherapy optimization is challenging. Most dose escalation studies assumed chronic hypoxia and we also consider hypoxia imaging and optimization for chronic hypoxia. While the many aspects of the mathematical methods described in this book can are universal and be applied to both chronic and acute hypoxia, acute hypoxia would require different approaches to address its dynamic nature [27].

1.3 3D Imaging of Tumor Hypoxia

1.3.1 PET Biomarkers of Hypoxia

As we already mentioned, tumor hypoxia has been historically measured using invasive electrode measurements [7] but significant progress in this area has been achieved using functional imaging with PET and MRI [8, 10, 12]. Several PET biomarkers have been developed to characterize chronic tumor hypoxia regions with low pO_2. We present an overview and comparison of some PET hypoxia biomarkers that are relatively widely used: ^{18}F-Fluoromisonidazole (^{18}F-FMISO) [8, 10, 11], ^{18}F-fluoroazomycin arabinoside (^{18}F-FAZA) [10, 13, 28], and Cu-diacetyl-bis(N^4)-methyl thiosemicarbazone (Cu-ATSM) [10, 29, 30].

(a) **^{18}F-Fluoromisonidazole**

The synthesis of ^{18}F-FMISO was proposed and researched by the investigators at the University of Washington who subsequently demonstrated the feasibility of ^{18}F-FMISO PET for imaging the hypoxic fractions in a variety of human tumors [8, 11, 26]. The development of ^{18}F-FMISO stems from the discovery that some radio sensitizing drugs such as misonidazole selectively bounds to the molecules in viable hypoxic cells in vivo and in vitro. Many investigations have shown that ^{18}F-FMISO binds stably to the same population of hypoxic cells as misonidazole leading to a conclusion that ^{18}F-FMISO is useful as a hypoxia imaging agent in vivo with PET. The ^{18}F FMISO PET hypoxia imaging is based on the hypothesis that radio sensitizers such as misonidazole that are not consumed like O_2 and could diffuse farther than O_2 and radiosensitize the poorly perfused hypoxic cells. Positive accumulation of ^{18}F-FMISO is proportional to the extent of hypoxia in the pO_2 range of a few mmHg and the high ^{18}F-FMISO uptake implies a low tissue O_2 concentration. The sample images of hypoxia in glioblastoma patients measured by ^{18}F-FMISO PET are shown in Fig. 1.2.

The quantification of hypoxic regions in ^{18}F-FMISO studies is based on a venous blood sample drawn during the imaging sequence to calculate a tissue-blood ratio. The presence of hypoxia in a tumor pixel is defined by a tumor/blood (T/B) ^{18}F-FMISO ratio that is above a certain cut off value. Based on the previous data, this cut off value was set to T/B = 1.2 [8]. Currently, ^{18}F-FMISO continues to be one of the leading pharmaceuticals in PET imaging for the evaluation and quantification of tumor hypoxia. The ^{18}F-FMISO is less confounded by blood flow and its uptake after 2 h presents a measure of the inadequate regional oxygen partial pressure at the time of radiopharmaceutical administration.

(b) **^{18}F-fluoroazomycin Arabinoside**

The disadvantages of ^{18}F-FMISO include relatively slow clearance from blood and non-target normal tissues which results in low values of target-to-background

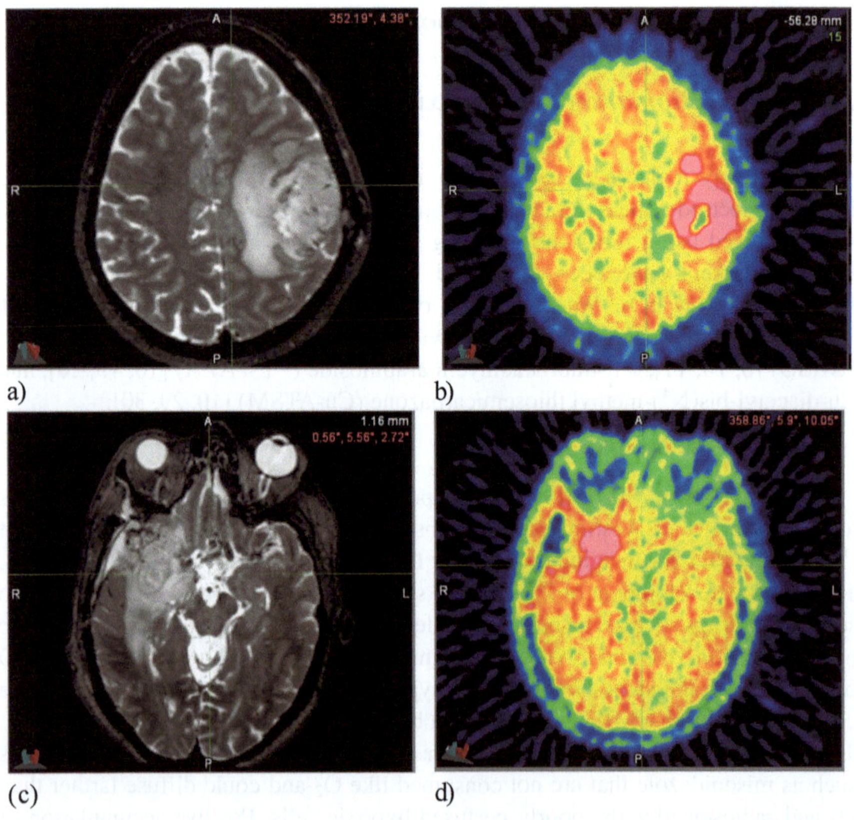

Fig. 1.2 T2 MRI (**a, c**) and ^{18}F-FMISO PET (**b, d**) images of glioblastoma for two different patients

ratios. The threshold value for target volume definition depends on the background level of the tracer in the normal tissue; therefore, tracers with relatively low retention in normal tissues would yield higher threshold values. To improve the target to-background ratios, the ^{18}F-FAZA hypoxia biomarker was synthesized and studied in various tumor-bearing mice using PET and autoradiography. The ^{18}F-FAZA binds to the intracellular macromolecules in viable cells at low oxygen tensions but compared with misonidazole the ^{18}F-FAZA displayed rapid clearance from blood, yielding more favorable T/B ratios [10, 28].

(c) **Cu-diacetyl-bis(N^4)-methylthiosemicarbazone**

The research on new hypoxia tracers has shown that the copper chelate Cu-ATSM can be an effective marker for hypoxic tissue; because it is selectively trapped in hypoxic tissue but rapidly washed out from well-oxygenated cells [10, 29, 30]. The application of Cu-ATSM in PET imaging is based on the positron emitting isotopes of copper (^{60}Cu, ^{61}Cu, ^{62}Cu, and ^{64}Cu). For example, to confirm in vivo that the degree of ^{60}Cu-ATSM accumulation is pO2 dependent, the uptake of ^{60}Cu-ATSM correlated with tumor pO2 was studied in two rat models. The tissue pO2 was measured by a needle oxygen electrode and tissue oxygen levels were manipulated in experiments [29, 30]. These data confirmed that ^{60}Cu-ATSM retention in tissue is pO2 dependent and greater retention is noted in hypoxic tissues. Additionally, the ^{60}Cu-ATSM is quickly washed out from well-oxygenated tissues.

1.3.2 Computing pO_2 from ^{18}F-FMISO PET Images

The distributions of physical dose and the OER are needed to compute the cellular response in radiotherapy target volumes. For precise characterization of oxygen effect, PET images of hypoxia that present a distribution of SUV (Standardized Uptake Value) needs to be converted to the distribution of pO_2 and then the pO_2 can be converted to the OER using Eqs. 1.5 or 1.6. The ^{18}F-FMISO PET images include a distribution of SUV; therefore, the SUV needs to be converted to pO_2 and then the pO_2 needs to be converted to the OER using Eq. 1.5 or 1.6.

The measurements and simulations show that accumulation of ^{18}F-FMISO is inversely related to pO_2 [30–33]. Based on these measurements, an empirical equation for dependence of the SUV on the pO_2 can be proposed. Assuming additionally the boundary conditions with the maximum SUV_{max} at hypoxia ($pO_2 = 0$ mmHg) and minimum SUV_{min} at aerobic conditions ($pO_2 = 100$ mmHg), we can derive the following generalized equation for dependence of ^{18}F-FMISO SUV on pO_2:

$$SUV(pO_2) = SUV_{min} + (SUV_{max} - SUV_{min})\exp(-RpO_2), \tag{1.9}$$

where R is a parameter. The measurements of ^{18}F-FMISO SUV as a function of pO_2 [31] are used for the estimation of parameter R in Eq. 1.9. Figure 1.3 shows that Eq. 1.9 with the parameter $R = 0.12$ describes the measured data. Using Eq. 1.9, we

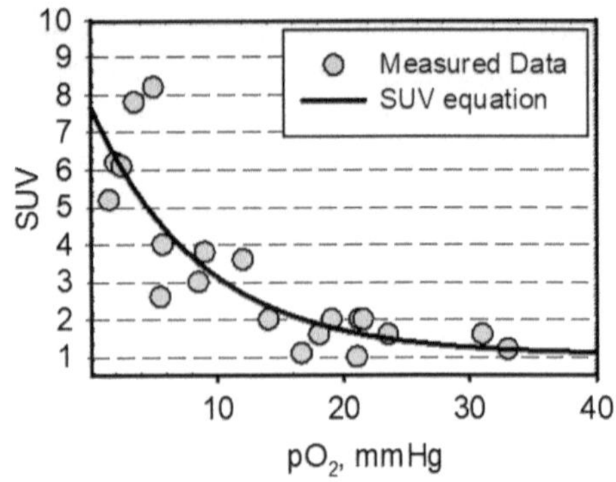

Fig. 1.3 Dependence of SUV on pO_2 for ^{18}F-FMISO PET hypoxia images [25]

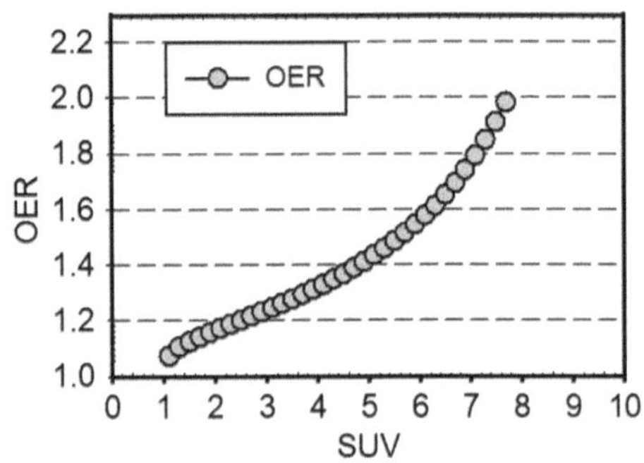

Fig. 1.4 Dependence of OER on SUV for ^{18}F-FMISO PET hypoxia images [25]

can show that pO_2 is given by:

$$pO_2 = \frac{1}{R} \ln \left\{ \frac{SUV(pO_2) - SUV_{min}}{SUV_{max} - SUV_{min}} \right\}, \tag{1.10}$$

We note that the dependence of the OER and the SUV on pO_2 is steeply hyperbolic as it is described by Eqs. 1.5 and 1.9 and shown in Figs, 1.1 and 1.3 respectively, However, the dependence of the OER on the SUV is smooth as shown in Fig. 1.4 and may be less prone to the uncertainties in the SUV. The algorithms to derive the OER from the SUV of the PET hypoxia biomarker Cu-ATSM have also been investigated and are very similar to the ^{18}F-FMISO-based algorithm [32, 33].

1.3.3 MRI Biomarkers of Hypoxia

The MRI based approaches for hypoxia measurements are minimally invasive and repeatable because they do not require ionizing radiation. We present an overview of Dynamic Contrast Enhanced (DCI) MRI and Blood Oxygen Level Dependent (BOLD) MRI [10].

(a) **Dynamic Contrast Enhanced (DCI) MRI**

The DCI MRI uses kinetic modeling of injected contrast reagents to characterize perfusion; therefore, it can yield information on the flow, volumes, and permeability of vessels [34]. The contrast agents generally contain chelates of gadolinium (Gd), which is an effective inducer of magnetic T1 relaxation and thus induces an increase in signal intensity under T1-weighted conditions. Alternative agents contain ferrous iron, an effective T2 agent, and thereby promote a decrease in signal intensity. It has been suggested that DCE-MRI may be used for minimally invasive pO2 measurements because there is biochemical and empirical relationship between vessel permeability and hypoxia [35]. It has been shown that the functional parameters derived from DCE-MRI correlate with tumor oxygenation in human cervix carcinomas [36]. Also, the high-risk tumor volumes derived from DCE-MRI have shown a correlation with radiotherapy response following radiotherapy of cervix cancers [12].

(b) **Blood Oxygen Level Dependent (BOLD)** [10] **MRI**

The BOLD MRI allows the visualization of time-dependent changes in near real time and is sensitive to the presence of changing vascular oxygen tension in the tissue. BOLD MRI is expected to provide a sensitive index of the oxygenation status of tissue immediately adjacent to perfused microvessels. Image contrast in BOLD MRI relies on the balance between the quantities of paramagnetic deoxyhemoglobin and diamagnetic oxyhemoglobin and the effects of deoxyhemoglobin on magnetic resonance (MR) signal. This technique does not require injection of contrast agent or ionizing radiation BOLD MRI does not directly measure pO2 or changes to pO2 in response to vasomodulation. Instead, it is sensitive to deoxyhemoglobin, which is present in perfused, deoxygenated tissues.

BOLD MRI produces quick readouts, allowing investigators to instantly monitor any changes in intravascular pO2 status. Deoxyhemoglobin serves as an intravascular contrast agent because it brings about an increase in the transverse relaxation rates of blood and surrounding tissues. As the oxygenation of hemoglobin is directly proportional to the pO2 of blood, which is assumed to be in equilibrium with tissue pO2.

1.3.4 Computing pO_2 from DCE MRI

We describe a method to derive pO_2 from the DCE MRI images proposed in [23]. In this method, it was assumed that the average concentration of tracer at a given point in the patient is proportional to the blood plasma concentration. As a high blood plasma concentration indicates a high pO_2, the latter may be taken to be proportional to the concentration of the tracer. For normalization of this method, it was assumed that the maximum global concentration of tracer happens at the points with maximum arterial pressure of oxygen that is typically 100 mm Hg. Therefore, the following equation for pO_2 was proposed:

$$pO_2 = \frac{\overline{C}}{\overline{C}_{\max}} pO_{2,\max}, \tag{1.11}$$

where $\overline{C}$ is the average concentration of tracer at a given point, $pO_{2,max}$ is the maximum arterial pressure and $\overline{C}_{max}$ is the maximum concentration of tracer that is expected to achieve at the maximum arterial pressure arterial pressure of oxygen.

The average concentration of tracer $\overline{C}$ is calculated using the following equation:

$$\overline{C} = \frac{1}{T}\int_0^T C(t)dt, \tag{1.12}$$

where T is the selected time. In general, this model assumes a linear dependence between the tracer concentration and the pO_2 in the regions between maximum and minimum traced uptake.

1.4 Cell Survival Models with Different Oxygen Levels

1.4.1 Liner-Quadratic Cell Survival Model

The simulations in this book assume that the liner-quadratic (LQ) model describes cell survival with acceptable accuracy at the dose levels below 10 Gy. The LQ model assumes that two components define cell killing by radiation - one of these components is proportional do the dose and second is proportional to the square of dose. The LQ cell survival model is given by

$$-\ln S(d) = \alpha d + \beta d^2, \tag{1.13}$$

where and α and β are the parameters of the LQ model, d is the radiation dose, and S(d) is the cell surviving fraction [1, 37]. The parameter α defines cell survival at low doses and parameter β defines the cell survival at larger doses.

The measurements of oxygen effect have shown that the cell survival curves are essentially the same shape with and without oxygen; therefore, the OER is approximately independent of the level of survival [22, 38]. The OER can be considered as a dose-modifying factor by which the dose in the presence of O_2 should be multiplied to give the survival curves for hypoxic conditions. However, as we already discussed in Sect. 1.2.2, there is a dependence of the OER on radiation dose that results in the smaller values of OER at lower doses and the larger values of OER at larger doses. To consider this dual radiation action, it is possible to introduce different values of the OER to consider dose dependence; for example, the OER_α for the linear component of (α) of radiation damage and the OER_β for the quadratic component (β) of radiation damage [21]. The LQ model with dose independent OER is given by:

$$\ln S(d) = -\alpha_{ox}\frac{d}{OER_\alpha} - \beta_{ox}\left(\frac{d}{OER_\beta}\right)^2, \quad (1.14)$$

where $\alpha_{ox,}$ and β_{ox} are the parameters of the LQ survival model for completely oxygenated cells.

The application of Eq. 1.14 would require application of Eq. 1.5 to each of the parameters OER_α and OER_β and would result in larger overall number of radiobiological parameters. Therefore, we use the following simplified equation for hypoxic tumors:

$$\ln S(d.OER)) = -\alpha_{ox}\frac{d}{OER} - \beta_{ox}\left(\frac{d}{OER}\right)^2, \quad (1.15)$$

where we assume that the OER = OER_α is smaller at low doses where parameter α dominates and the OER = OER_β is larger at higher doses where parameter β dominates.

1.4.2 Cell Survival Models at High Doses

The LQ model underestimates cell survival at the doses larger than 10 Gy where the dose response becomes essentially linear. There is an increased application of high doses in radiation therapy partially due to implementation of the Stereotactic Body Radiation Therapy (SBRT). The fractionated SBRT may include the doses up to 20 Gy per fraction and the single fraction SBRT treatments may use doses up to 34 Gy [39, 40]. The precise modeling of the tumor control probability (TCP) in hypofractionated radiotherapy may require the application of alternate cell survival models. Therefore, several hybrid cell survival models have been proposed that maintain properties of the LQ cell survival model at the low doses and then transition from the quadratic function of dose to a linear behavior as the dose fraction size becomes larger [40–43].

To describe the response of oxygenated cells at high doses in our simulations in this book, we apply the Universal Survival Curve (USC) which uses the LQ model at lower doses and the multitarget model at higher doses [40]. The cell survival fraction is a function of dose d in the USC is given by

$$-\ln S(d) = \begin{cases} \alpha d + \beta d^2, & d \le D_T \\ \frac{1}{D_0}d - \frac{D_q}{D_0}, & d > D_T \end{cases}, \quad (1.16)$$

where α, β, D_T, D_0, and D_q are the parameters of the model. A relationship between the parameters of the quadratic and linear components was imposed to provide a continuity of cell survival curve and its derivative at the junction point.

$$\beta = \frac{(1 - \alpha D_0)^2}{4 D_0 D_q}, \tag{1.17}$$

$$D_T = \frac{2 D_q}{1 - \alpha D_0}. \tag{1.18}$$

An alternative formulation of the USC that uses only three parameters α, β and D_T was proposed [41]:

$$- \ln S(d) = \begin{cases} \alpha d + \beta d^2, & d \le D_T \\ \alpha D_T + \beta D_T^2 + s_{\max}(d - D_T), & d > D_T \end{cases}, \tag{1.19}$$

where

$$s = \alpha + 2\beta D_T. \tag{1.20}$$

A comparison of the USC in these different formulations in shown in Fig. 1.5a with the parameters α, β, and D_T for non-small cell lung cancer described in Chap. 3. This comparison illustrates the equivalency of these two approaches.

Hypoxic cells are more radioresistant than oxygenated cells and the effective dose for inactivation of hypoxic cell can be computed using the OER [1, 21]:

$$d_{ox} = \frac{d_{hyp}}{OER}. \tag{1.21}$$

Substituting Eq. 1.22 in Eq. 1.19, we obtain the cell survival curve for hypoxic cells is given by [44]:

$$- \ln S(d, OER) = \begin{cases} \alpha_{ox} \frac{d}{OER} + \beta_{ox} \left(\frac{d}{OER}\right)^2, & d \le OER \Delta D_T \\ \alpha_{ox} D_T + \beta_{ox} D_T^2 + s_{\max}\left(\frac{d}{OER} - D_T\right), & d > OER \Delta D_T \end{cases} \tag{1.22}$$

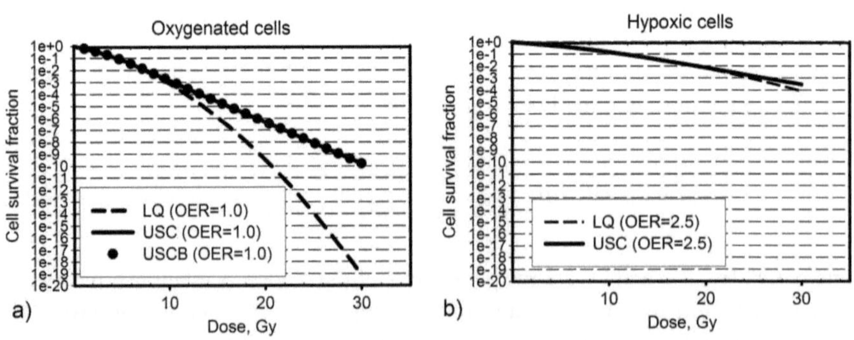

Fig. 1.5 Comparison of LQ and USC models for oxygenated (**a**) and hypoxic (**b**) cells [44]. The USC model is computed using initial (USC [40]) and modified (USCB [41]) equations

Equation 1.22 predicts that hypoxia effectively extends the applicability of the LQ model to higher doses by the factor of OER. The maximum value of the OER for the dose levels used in hypofractionated radiotherapy is estimated as $OER_{max} = 3$ [1]. The parameter D_T in the USC for non-small cell lung cancer is estimated as 6.2 Gy (see Chapter 3), therefore, the application of the LQ model for hypoxic cell is feasible below 18.6 Gy. There are also estimates that the LQ model can be used for the doses up to 10 Gy for oxygenated cells [45] that would be converted to 30 Gy under hypoxic conditions using Eqs. 1.21 and 1.22. A comparison of the LQ and USC models for oxygenated and hypoxic cells with OER = 2.5 is shown in Fig. 1.5(b).

1.5 Tumor Control Probability

1.5.1 TCP in Binomial and Exponential Forms

(a) **TCP in Binomial Form**

The equation of the TCP in cancer radiotherapy stems in part from the multiplication rule for the probabilities of independent random events. To derive this equation, we assume that an irradiated tumor has *N* initial clonogenic cells with an independent capability to initiate tumor regrowth. We also assume that each clonogenic cell has the same survival probability *S* given by the average surviving fraction for the radiotherapy treatment. In general, the average cell surviving fraction depends on the total dose, dose fractionation, overall treatment time, and several other radiobiological parameters that reflect the rate of tumor-cell repopulation, intrinsic radiation resistance, hypoxia, and other radiobiological factors.

The multiplication rule for events with independent probabilities states that the probability that several events occur is found by multiplying the probabilities of each event. The probability of not initiating tumor regrowth for each clonogenic cell is given by *1-S* where *S* is survival probability; therefore, the TCP can be computed using the multiplication rule as [46]:

$$TCP = (1 - S)^N. \tag{1.23}$$

The TCP decreases if *N* increases, and this effect constitutes the nature of the volume effects in the TCP because *N* increases with tumor volumes. The rate of the decrease is larger for larger *S*, for example, in the hypoxic regions and for lower doses.

(b) **TCP in Exponential Form**

It can be shown that Eq. 1.23 can be approximated for large N and small S by the following equation [47–50]:

$$TCP = \exp(-NS) \tag{1.24}$$

with an approximation error bounded by NS^2 that is small for small S [51]. We note that Eq. 1.24 is also related to Poisson probability distribution and will be used in the simulation in this book because of its one-parametric nature.

As an example of the homogeneous TCP model, we consider a radiotherapy model with n dose fractions, fractional dose d and the exponential proliferation model for clonogenic cells. We also assume that the tumor is homogeneous with radiosensitivity described by the cell surviving fraction S given by Eq. 1.13. In this case the TCP can be written as

$$TCP(N, \alpha, \beta, \lambda) = \exp\Big(-N\big\{\exp\{-\alpha d - \beta d^2\}\big\}^n \exp(\lambda T)\Big), \tag{1.25}$$

where T is the total duration of radiotherapy and λ is the repopulation rate for clonogenic cells. The TCP formulation given by Eq. (1.25) assumes a uniform population of tumors with identical numbers of clonogenic cells, radiosensitivity and repopulation kinetics. This form proves to be a reasonable fit to clinical datasets and has proved valuable in clinical radiobiology despite the non-uniform distributions of parameters.

1.5.2 *Density of Clonogenic Cells*

The total number of clonogenic cells N is a key parameter that is required for computing the TCP. For practical calculations, it is convenient to change from the total number N of clonogenic cells to directly measurable parameters such as tumor volume V and density of clonogenic cells ρ. To convert from the number of clonogenic cells N to the tumor volume and density of clonogenic cells, the following power-law relationship between the initial tumor volume and the number of clonogenic cells was suggested [46, 52, 53]:

$$N = \rho V^a, \tag{1.26}$$

where ρ is the density of clonogenic cells, V is the initial tumor volume and a is a volume exponent. A linear or nearly linear relationship between the initial number of clonogenic cells and the tumor volume was previously obtained using measured TCP data and Eq. 1.26. The different values of the volume exponent in the range between 0.37 and 1.0 have been reported based on the analysis of clinical data. The value of $a = 1$ assumes that the number of clonogenic cells is directly proportional to the initial tumor volume V.

Volumetric densities of clonogenic cells that were estimated from clinical data by several authors and, also, using clonogenic fractions from plating efficiencies of in vitro human tumors are summarized in [54]. For instance, for head and neck cancers, most estimates for the density of clonogenic cells ρ are in the range from 1.0E + 6 to 1.0E + 7 clonogenic cells per cm^3. Assuming an effective (overall) cell density of 1.0E + 8 cells per cm^3, a representative density of clonogenic cells of 1.0E + 6 cells per cm^3 corresponds to a fraction of 1% for clonogenic cells.

1.5.3 Heterogeneous TCP Model

Equations 1.23–1.26 assume the uniform distribution of cell surviving fraction S and density of clonogenic cells ρ within the tumor. These distributions are not uniform in human tumors. To derive the TCP equations for variable tumor radiosensitivity and density of clonogenic cells, we assume that a heterogeneous tumor consists of K sub-volumes $V_k, k = \overline{1, K}$ with the corresponding densities of clonogenic cells and the cell surviving fractions $(\rho_k, S_k), k = \overline{1, K}$, The multiplication rule for the independent probabilities leads to the following TCP equation in binomial form for this heterogeneous tumor model:

$$TCP = \prod_{k=1}^{K} (1 - S_k)^{\rho_k V_k}, \tag{1.27}$$

that can be written in the exponential form for small values of cell surviving fractions S_k:

$$TCP = \exp\left(- \sum_{k=1}^{K} \rho_k V_k S_k \right). \tag{1.28}$$

For practical applications, it is useful to define a parameter called the tumor-average cell surviving fraction given by:

$$S_{TA} = \frac{1}{\rho_{TA}} \sum_{k=1}^{K} \rho_k v_k S_k, \tag{1.29}$$

where $v_k = V_k/V_{tot}$ is the fractional tumor sub-volume defined by the total tumor volume V_{tot} and the tumor-average density of clonogenic cells ρ_{TA} is given by

$$\rho_{TA} = \sum_{k=1}^{K} \rho_k v_k. \tag{1.30}$$

Using parameters defined in Eqs. 1.29 and 1.30, the TCP of a tumor with the variable radiosensitivity and density of clonogenic cells can be written in the compact form as:

$$TCP = \exp(-\rho_{TA} V_{tot} S_{TA}). \tag{1.31}$$

1.5.4 Population Based TCP Models

In the case of a heterogeneous population of tumors, the heterogeneous TCP can be obtained by integrating Eq. 1.25 over the probability distribution of the relevant parameters. For instance, we can consider three major sources of heterogeneity: number of clonogenic cells N, radiosensitivity parameter α and repopulation rate λ. Assuming normal distribution of parameters for simplicity we obtain the following equation for heterogeneous TCP model [55, 56]:

$$TCP_{het} = \int_{-\infty}^{\infty} TCP(N, \alpha, \beta, \lambda) F(N, \alpha, \lambda) dN d\alpha d\lambda, \tag{1.32}$$

where the probability density function F(N,α,β) is given by:

$$F(N, \alpha, \lambda) = \frac{1}{(2\pi)^{3/2} \sigma_N \sigma_\alpha \sigma_\lambda} \exp\left\{ \frac{\left(N - \overline{N}\right)^2}{2\sigma_N^2} \right\} \exp\left\{ \frac{(\alpha - \overline{\alpha})^2}{2\sigma_\alpha^2} \right\} \exp\left\{ \frac{\left(\lambda - \overline{\lambda}\right)^2}{2\sigma_\lambda^2} \right\} \tag{1.33}$$

1.6 Normal Tissue Complication Probability

The efficacy of radiation therapy is defined not only by the TCP but also by the normal tissue complication probability (NTCP). These two parameters define the therapeutic ratio of radiotherapy. The clinical data demonstrate the NTCP exhibits dependence on the radiation dose and irradiated volume; however, it is more difficult to relate cell death to the biological effect in the way it was done in mechanistical models for the TCP. Therefore, most of the NTCP models are phenomenological fit-to-data equations that describe the clinical data in a most optimal way [57, 58].

As an example, we describe the Lyman-Kutcher-Burman (LKB) model that is currently the most used NTCP model [59, 60]. The LKB model can be expressed as:

$$NTCP = \frac{1}{(2\pi)^{1/2}} \int_{-\infty}^{t} \exp\left\{ -\frac{x^2}{2} \right\} dx, \tag{1.34}$$

where

$$t = \frac{EUD - TD_{50}}{mTD_{50}}. \tag{1.35}$$

The following notations are used in Eqs. 1.34 and 1.35: TD_{50} is the uniform dose given to the entire organ volume that results in 50% risk of complications, m is a measure of the slope of the sigmoid curve obtained from the integration of the normal distribution, and EUD is the equivalent uniform dose.

The EUD is the dose that if given uniformly to the entire organ volume will lead to the same NTCP as a nonuniform dose distribution. The EUD is computed by:

$$EUD = \left\{ \sum_{i=1}^{I} v_i D_i^{1/n} \right\}^n, \tag{1.36}$$

where v_i and D_i are the fractional volume and dose of each sub-volume used to tabulate the nonuniform dose distribution and n is the parameter that describes the magnitude of the volume effect. The volume effect is not explicitly present in the LKB model; however, it is implicitly present in the model using the EUD formalism.

1.7 Uncertainties in Radiobiological Models

The measured radiobiological parameters, for example, the parameters "alpha" and "beta" in the LQ cell survival curve are usually associated with large errors that may produce related errors in the radiobiological TCP and NTCP modeling. The inverse problem of fitting the TCP and NTCP models to clinical data to evaluate radiobiological parameters will result in certain uncertainties in radiobiological parameters. The radiobiological models are usually complex nonlinear functions; however, the uncertainties in measured radiobiological parameters and radiobiological functions can be related using the sensitivity functions and the linear perturbation theory.

The sensitivity function can be defined as [61]

$$s(x) = \frac{\delta F(x)/F(x)}{\delta x/x}, \tag{1.37}$$

where $\delta F(x)$ is the variation of radiobiological function $F(x)$ and δx is the variation of the radiobiological parameter x. For example, the sensitivity function $s(t) = 1.0$ means that the 1% variation in the radiobiological parameter will produce 1% variation in the radiobiological function. Sensitivity functions can be evaluated using direct perturbation methods and the related errors can be computed using the linear perturbation theory. In the direct perturbation method, a small error which is usually 1% is introduced into parameter x and the radiobiological function $F(x)$ is recalculated to evaluate the related variation $\delta F(x)$.

Assuming the tumor-volume has a constant systematic error of ΔV, the relative error of reconstructed radiobiological parameter Δx can be evaluated as:

$$\Delta F(x) = s(x)\Delta x, \tag{1.38}$$

Another useful application of sensitivity analysis is the comparative evaluation of the relative importance of several radiobiological parameters.

References

1. Hall EJ, Giaccia AJ. Radiobiology for the Radiologist. 6th ed. Philadelphia, PA: Lippincott Williams & Wilkins; 2006.
2. Bjork-Eriksson T, West C, Karlsson E, Mercke C. Tumor radiosensitivity (SF2) is a prognostic factor for local control in head and neck cancers. Int J Radiat Oncol Biol Phys. 2000; 46:13–19.
3. Buffa FM, Davidson SE, Hunter RD, Nahum AE, West CM. Incorporating biologic measurements (SF(2), CFE) into a tumor control probability model increases their prognostic significance: a study in cervical carcinoma treated with radiation therapy. Int J Radiat Oncol Biol Phys. 2001; 50(5):1113–1122.
4. Eschrich SA, Pramana J, Zhang H, Zhao H, Boulware D, Lee JH, Bloom G, Rocha-Lima C, Kelley S, Calvin DP, Yeatman TJ, Begg AC, Torres-Roca JF. A gene expression model of intrinsic tumor radiosensitivity: prediction of response and prognosis after chemoradiation. Int J Radiat Oncol Biol Phys. 2009; 75:489–496.
5. Begg AC. Predicting response to radiotherapy: evolutions and revolutions. Int J Radiat Biol. 2009; 85:825–836.
6. Spence AM, Muzi M, Link JM, O'Sullivan F, Eary JF, Hoffman JM, Shankar LK, Krohn KA. NCI-sponsored trial for the evaluation of safety and preliminary efficacy of 3'-deoxy-3'-[18F] fluorothymidine (FLT) as a marker of proliferation in patients with recurrent gliomas: preliminary efficacy studies. Mol Imaging Biol. 2009; 11(5):343–55.
7. Nordsmark M, Overgaard M, Overgaard J. Pretreatment oxygenation predicts radiation response in advanced squamous cell carcinoma of the head and neck. Radiother Oncol. 1996; 41: 31–39.
8. Spence AM, Muzi M, Swanson KR, O'Sullivan F, Rockhill JK, Rajendran JG, Adamsen TC, Link JM, Swanson PE, Yagle KJ, Rostomily RC, Silbergeld DL, Krohn KA. Regional hypoxia in glioblastoma multiforme quantified with [18F] fluoromisonidazole positron emission tomography before radiotherapy: correlation with time to progression and survival. Clin Cancer Res. 2008; 14:2623–30.
9. Rengan R, Rosenzweig KE, Venkatraman E, Koutcher LA, Fox JL, Nayak R, Amols H, Yorke E, Jackson A, Ling CC, Leibel SA. Improved local control with higher doses of radiation in large-volume stage III non-small-cell lung cancer. Int J Radiat Oncol Biol Phys. 2004; 60:741–747.
10. Tatum JL, Kelloff GJ, Gillies RJ, Arbeit JM, Brown JM, Chao KS, Chapman JD, Eckelman WC, Fyles AW, Giaccia AJ, Hill RP, Koch CJ, Krishna MC, Krohn KA, Lewis JS, Mason RP, Melillo G, Padhani AR, Powis G, Rajendran JG, Reba R, Robinson SP, Semenza GL, Swartz HM, Vaupel P, Yang D, Croft B, Hoffman J, Liu G, Stone H, Sullivan D. Hypoxia: importance in tumor biology, noninvasive measurement by imaging, and value of its measurement in the management of cancer therapy. Int J Radiat Biol. 2006; 82(10):699–757.
11. Rajendran JG, Schwartz DL, O'Sullivan J, Peterson LM, Ng P, Scharnhorst J, Grierson JR, Krohn KA. Tumor hypoxia imaging with [F-18] fluoromisonidazole positron emission tomography in head and neck cancer. Clin Cancer Res. 2006; 12(18): 5435–41.

12. Mayr NA, Huang Z, Wang JZ, Lo SS, Fan JM, Grecula JC, Sammet S, Sammet CL, Jia G, Zhang J, Knopp MV, Yuh WT. Characterizing tumor heterogeneity with functional imaging and quantifying high-risk tumor volume for earlier prediction of treatment outcome: cervical cancer as a model. Int J Radiat Oncol Biol Phys. 2012; 83(3):972–979.
13. Graves EE, Hicks RJ, Binns D, Bressel M, Le QT, Peters L, Young RJ, Rischin D. Quantitative and qualitative analysis of [(18)F]FDG and [(18)F]FAZA positron emission tomography of head and neck cancers and associations with HPV status and treatment outcome. Eur J Nucl Med Mol Imaging. 2016; 43(4):617–25.
14. Sövik A, Malinen E, Olsen D. Strategies for biologic image guided dose escalation: a review. Int J Radiat Oncol Biol Phys. 2009, 73: 650–658.
15. Das SK, Ten Haken RK, Functional and molecular image guidance in radiotherapy treatment planning optimization, Semin Radiat Oncol. 2011, 21(2): 111–118.
16. Salem A. Hypoxia-Targeted Dose Painting in Radiotherapy. Semin Radiat Oncol. 2023, 33: 298–306.
17. Mirzakhanian L, Bassalow R, Zaks D, Huntzinger C, Seuntjens J. IAEA-AAPM TRS-483-based reference dosimetry of the new RefleXion biology-guided radiotherapy (BgRT) machine. Med Phys. 2021; 48 (4):1884–1892.
18. Rajendran JG, Mankoff DA, O'Sullivan F, Peterson LM, Schwartz DL, Conrad EU, Spence AM, Muzi M, Farwell DG, Krohn KA. Hypoxia and glucose metabolism in malignant tumors: Evaluation by [^{18}F] Fluoromisonidazole and [^{18}F] Flurodeoxyglucose positron emission tomography imaging. Clinical Cancer Res 2004; 10: 2245–52.
19. Pugachev A, Shutlan R, Carlin S, Larson SM, Campa J, Ling CC, Humm JL. Dependence of FDG uptake on tumor microenvironment. Int J Radiat Oncol Biol Phys. 2005; 62(2):545–553.
20. Alper T, Howard-Flanders P. The role of oxygen in modifying the radiosensitivity of E. coli B. Nature, 1956; 178: 978–979
21. Wouters BG, Brown JM. Cells at intermediate oxygen levels can be more important than the "hypoxic fraction" in determining tumor response to fractionated radiotherapy Rad. Research. 1997, 147: 541–550.
22. Carlson DJ, Stewart RD, Semenenko VA. Effects of oxygen on intrinsic radiation sensitivity: A test of the relationship between aerobic and hypoxic linear-quadratic LQ model parameters. Med Phys. 2006; 33(9):3105–3115.
23. Malinen E, Sovik A, Hristov D, Bruland ØS, Olsen DR. Adapting radiotherapy to hypoxic tumors. Phys Med Biol. 2006; 51: 4903–4921.
24. Sovik A, Malinen E, Bruland OS, Bentzen SM, Olsen DR, Optimization of tumor control probability in hypoxic tumors by radiation dose redistribution. Phys Med Biol. 2007; 52: 499–513.
25. Chvetsov AV, Muzi M. Equivalent uniform aerobic dose in radiotherapy for hypoxic tumors. Phys Med Biol. 2024; 69: 085011.
26. Rajendran JG, Krohn KA. F-18 fluromisonidazole for imaging tumor hypoxia: imaging the microenvironment for personalized cancer therapy. Semin Nucl Med. 2015; 45: 151–162.
27. Taylor E. A simple mathematical model of cyclic hypoxia and its impact on hypofractioated radiotherapy. Med Phys. 2023;50(3):1893–1904.
28. Piert M, Machulla H-J, Picchio M, Reischl G. Ziegler S, Kumar P, Wester H-J, Beck R, McEwan AJB, Wiebec LI, Schwaiger M. Hypoxia-Specific Tumor Imaging with 18F-Fluoroazomycin Arabinoside. J Nucl Med. 2005; 46(1) 106–113.
29. Chao KS, Bosch WR, Mutic S, Lewis JS, Dehdashti F, Mintun MA, Dempsey JF, Perez CA, Purdy JA, Welch MJ. A novel approach to overcome hypoxic tumor resistance: Cu-ATSM-guided intensity-modulated radiation therapy. Int J Radiat Oncol Biol Phys. 2001;49(4): 1171-1182.
30. Lewis JS, McCarthy DW, McCarthy TJ, Fujibayashi Y, Welch MJ. Evaluation of 64Cu-ATSM in Vitro and in Vivo in a Hypoxic Tumor model. J Nucl Med. 1999; 40: 177–183.
31. Piert M, Machulla H-J, Becker G, Aldinger P, Winter E, Bares R. Dependency of the [^{18}F] Fluromisonidasole uptake on oxygen delivery and tissue oxygenation in the porcine liver. Nucl Med Biol. 2000; 27: 693–700.

32. Titz B, Jeraj R. An imaging-based tumor growth and treatment response model: investigating the effect of tumor oxygenation on radiation therapy response. Phys Med Biol. 2008; 53: 4471–4488.
33. Bowen SR, van der Kogel AJ, Nordsmark M, Bentzen SM, Jeraj R. Characterization of positron emission tomography hypoxia tracer uptake and tissue oxygenation via electrochemical modeling. Nucl Med Biol. 2011; 38: 771–780.
34. Tofts PS, Brix G, Buckley DL, Evelhoch JL, Henderson E, Knopp MV, Larsson HB, Lee TY, Mayr NA, Parker GJ, Port RE, Taylor J, Weisskoff RM. Estimating kinetic parameters from dynamic contrast-enhanced T(1)-weighted MRI of a diffusible tracer: standardized quantities and symbols. J Mag Reson Imaging. 1999; 10:223–232.
35. Foo SS, Abbott DF, Lawrentschuk N. Scott AM. Functional imaging of intratumoral hypoxia. Mol. Imaging Biol. 2004; 6: 291–305.
36. Lyng H, Vorren A O, Sundfor K, Taksdal I, Lien HH, Kaalhus A, Rofstad E K. Assessment of tumor oxygenation in human cervical carcinoma by use of dynamic Gd-DTPA-enhanced MR imaging. J Magn Reson Imaging. 2001; 14: 750–756.
37. Fowler JF. The linear-quadratic formula and progress in fractionated radiotherapy. Br. J. Radiol. 1989; **62:** 679–94.
38. Johns HE, Cunningham JR. The Physics of Radiology. 4th ed. Springfield, Illinois. Charles C. Thomas, Publisher; 1983.
39. Chi A, Liao Z, Nguyen NP, Xu J, Stea B, Komaki R. Systemic review of the patterns of failure following stereotactic body radiation therapy in earlier-stage non-small-cell lung cancer: Clinical implications. Radiother Oncol. 2010; 94: 1–11.
40. Park C, Papiez L, Zhang S, Story M, Timmerman RD. Universal survival curve and single fraction equivalent dose: useful tools in understanding potency of ablative radiotherapy. Int J Radiat Oncol Biol Phys. 2008; 70: 847–852.
41. Beuve M, Alphonse G, Maalouf M, Colliaux A, Battiston-Montagne P, Jalade P, Balanzat E, Demeyer A, Bajard M, Rodriguez-Lafrasse C. Radiobiologic parameters and local effect model predictions for head-and-neck squamous cell carcinomas exposed to high linear energy transfer ions. Int J Radiat Oncol Biol Phys. 2008; 71: 635–642.
42. Guerrero M, Carlone M Mechanistic formulation of a lineal-quadratic-linear (LQL) model: split-dose experiments and exponentially decaying sources Med Phys. 2010; 37: 4173–4181.
43. Liu F, Tai A, Lee P, Biswas T, Ding GX, El Naqa I, Grimm J, Jackson A, Kong FS, LaCouture T, Loo B Jr, Miften M, Solberg T, Li XA. Tumor control probability modeling for stereotactic body radiation therapy of early-stage lung cancer using multiple bio-physical models. Radiother Oncol. 2017; 122: 286–294.
44. Chvetsov AV, Hanin LG, Stewart RD, Zeng J, Rengan R, Lo SS. Tumor control probability in hypofractionated radiotherapy as a function of total and hypoxic tumor volumes. Phys Med Biol. 2021; 66: 125010.
45. Brenner DJ. The linear-quadratic model is an appropriate methodology for determining isoeffective doses at large doses per fraction. Semin Radiat Oncol. 2008; 18: 234-239.
46. Brenner DJ. Dose, volume, and tumor-control predictions in radiotherapy Int J Radiat Oncol Biol Phys. 1993; 26: 171–179.
47. Munro TR, Gilbert CW. The relation between lethal doses and radiosensitivity of tumor cells. Br. J. Radiol. 1961; 34: 246–251.
48. Porter EH. The statistics of dose/cure relationships for irradiated tumors. Part I. Br J Radiol. 1980; 53: 210–227.
49. Porter EH. The statistic of dose\cure relationships for irradiated tumor. Part II Br J Radiol. 1980; 53: 336–345.
50. Zaider M, Hanin L. Tumor Control Probability in radiation treatment. Med Phys. 2011; 38: 574–583.
51. Taylor HM, Karlin S. An Introduction to Stochastic Modeling. San Diego, CA. Academic Press; 1998.
52. Bentzen SM, Overgaard J, Thames HD, Overgaard M, Hansen PV, Maase H, Meder J. Clinical radiobiology of malignant melanoma. Radioth Oncol. 1989; 16(3):169–182.

53. Bentzen SM, Thames HD. Tumor volume and local control probability: Clinical data and radiobiological interpretations. Int J Radiat Oncol Biol Phys. 1996; 36:247–251.
54. Wigg DR. Applied Radiobiology and Bioeffect Planning. Madison, WI. Medical Physics Publishing; 2001.
55. Webb S, Nahum AE. A model for calculating tumor control probability in radiotherapy including the effects of inhomogeneous distributions of dose and clonogenic cell density. Phys Med Biol. 1993; 38:653–666.
56. Roberts SA, Hendry JH. A realistic closed-form radiobiological model of clinical tumor-control data incorporating inter-tumor heterogeneity. Int J Radiat Oncol Biol Phys. 1998; 41: 689–699.
57. Moiseenko V, Deasy JO, Van DyK J. Radiobiological modeling for treatment planning. In: The modern technology of radiation oncology. A compendium for medical physicists and radiation oncologists. Vol. 2. Editor Van Dyk J. Madison, Wisconsin. Medical Physics Publishing; 2005.
58. Bentzen SM, Constine LS, Deasy JO, Eisbruch A, Jackson A, Marks LB, Ten Haken RK, Yorke ED. Quantitative Analyses of Normal Tissue Effects in the Clinic (QUANTEC): an introduction to the scientific issues. Int J Radiat Oncol Biol Phys. 2010;76(3 Suppl), S3–9.
59. Lyman JT, Wolbarst AB. Optimization of radiation therapy, IV: A dose-volume histogram reduction algorithm. Int J Radiat Oncol Biol Phys. 1989; 17(2): 433–436
60. Kutcher GJ, Burman C. Calculation of complication probability factors for non-uniform normal tissue irradiation: The effective volume method. Int J Radiat Oncol Biol Phys. 1989; 16: 1623–1630.
61. Chvetsov AV. Equivalent uniform RBE-weighted dose in eye plaque brachytherapy Med. Phys. 2024; 51(4): 3093–3100.

Chapter 2
Tumor Models for Optimization of Radiotherapy Response

2.1 Characteristics of Tumor Response Models

Models of tumor response to radiation therapy are needed for evaluation of the TCP, development of new dose fractionation schedules and other improved radiotherapy approaches. Modelling tumor response to radiation therapy has been investigated by many authors and many radiobiological tumor-response models have been proposed [1–13]. These models span from the simplest 2-level cell-population tumor response models [3–8] to the more complicated computer implementations that are based on 3D individual cell simulations using random processes simulated by Monte Carlo methods [11–13]. The major challenge in this research is to design robust tumor-response models that are complex enough to describe the underlying radiobiological processes of tumor response, and mathematically robust to provide a reliable parameter estimation using measured radiobiological, imaging, and other clinical data.

Validation using high accuracy measured data is an important component in the development of any radiobiological model. Complex models of tumor response are known [9–13], but they often contain a significant number of parameters. Too many parameters in the tumor response model can render the inverse reconstruction underdetermined, especially in the presence of several exponent functions and noisy input imaging data [14, 15]. A parsimonious model more likely outperforms complex models in the reconstruction of tumor response parameters from noisy measured data.

One of the simplest tumor response models with tumor hypoxia was proposed by Fisher for computational optimization of fractionation schedules [1, 2]. In this original work, Fisher used a model that assumes that a tumor consists of four cell subpopulations: viable and lethally damaged oxygenated cells and viable and lethally damaged hypoxic cells. The hypoxic sub-subpopulations are important for the definition of heterogeneous tumor radiosensitivity. The 4-level tumor model includes the main radiobiological determinants of tumor response to radiotherapy

A. V. Chvetsov, *Optimization Problems in Radiotherapy for Hypoxic Tumors*, Biological and Medical Physics, Biomedical Engineering,
https://doi.org/10.1007/978-981-96-7001-7_2

that include intrinsic cellular radiosensitivity, oxygen status, and proliferative potential. The 4-level approach was a major advancement in tumor-response analysis because it includes hypoxic cells and became a valuable tool for optimization of radiotherapy response. The 4-level cell population tumors model describes the total time-dependent number of cells in the tumor during fractionated radiotherapy; thus, we believe that this model can be used for comparison with imaging based and other measurements.

Fischer formulated the theory of splitting the entire tumor cell population into feasible sub-levels; however, he was not able to derive workable equations for the 4-level tumor model due to incorrect assumptions in the re-oxygenation model. In this chapter, we derive equations for the 4-level tumor model based on an optimal balance between complexity of underlying processes and minimizing the number of adjustable parameters to make a robust fit to the measured tumor-volume variation curves. The improved 4-level tumor model has been proposed for potential application for tumor-response analysis and quantitative analysis of measured tumor-volume variation curves during radiotherapy [16].

The 4-level tumor model is one of the most complete tumor-response models; therefore, it is a valuable tool for computer simulations but may be too complex for the analysis of measured data. To address this problem, different mathematical modifications of a 2-level cell-population tumor model with exponential clearance of lethally damaged cells have been proposed [3–8]. The 2-level tumor model separates the entire tumor cell population into two sub-populations of viable and lethally damaged cells and can be considered a special case of the 4-level tumor model for completely oxygenated tumors. One of the first implementations of this model in an analytical form was used for the evaluation of tumor radiosensitivity using tumor-volume variation reconstructed from two orthogonal two-dimensional (2D) projections [3]. This approach included significant volume uncertainties in the estimation of tumor volume.

The three-dimensional (3D) anatomical imaging techniques developed later in radiation therapy allowed an evaluation of tumor volume with higher accuracy [17, 18] and initiated renewed interest in this approach for parameter estimation. For instance, serial CT imaging of a shrinking tumor during radiotherapy for non-small cell lung cancer is shown in Fig. 2.1. Different mathematical modifications of the 2-level tumor model have been used with serial anatomical CT and MRI imaging. Recent work in cervical cancer studies with serial MRI provided a correlation of reconstructed cell survival fractions with measured tumor hypoxic fractions [4] and treatment outcomes [5]. Good agreement with the in vitro cell survival measurements was found for cell surviving fractions reconstructed from serial CT images obtained for lung and head-and-neck cancer [6, 7]. Most recently, a new application of the 2-level tumor model for optimization of fractionation schedules has been provided [8]. These studies suggest that the 2-level tumor model is potentially a valid instrument for tumor response assessment.

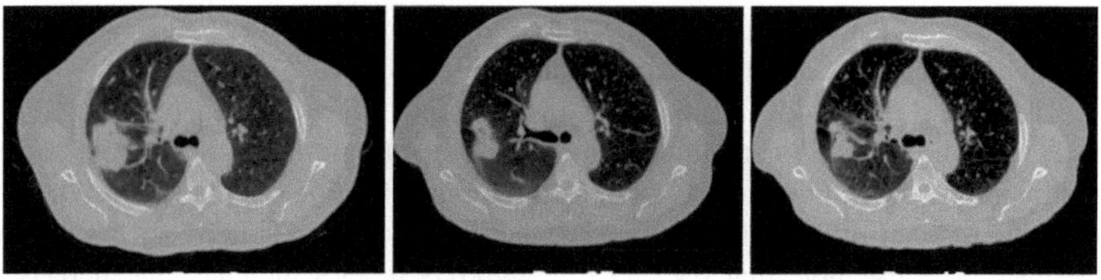

Fig. 2.1 Serial CT images of a shrinking tumor aquired during radiotherapy for non-small cell lung cancer on **a** day 2, **b** day 27, and **c** day 49

2.2 Four-Level Tumor Response Model

2.2.1 Four Sub-populations of Tumor Cells

The 4-level cell-population tumor response model was developed for optimization of dose fractionation schedules in radiation therapy and can be applied for evaluation of treatment response assessment, the TCP and quantitative analysis of measured tumor-volume variation curves during radiotherapy. The 4-level cell-population tumor response model will be called below simply as "4-level tumor model". The 4-level model assumes that a tumor consists of 4 cell subpopulations: viable and lethally damaged oxygenated cells and viable and lethally damaged hypoxic cells [1.2]. We define the four sub-populations of tumors cells as:

n_1—oxygenated viable cells,
n_2—oxygenated lethally damaged cells,
n_3—hypoxic viable cells, and.
n_4—hypoxic lethally damaged cells.

The total number of tumor cells N in the tumor is given by:

$$N = \sum_{i=1}^{4} n_i. \tag{2.1}$$

The lethally damaged cells exist in some intact form until they disintegrate, and their debris are removed from the tumor; therefore, they contribute to the bulk of the tumor even though they are unable to proliferate. The lethally damaged cells contribute to tumor volume until they disintegrate and are removed from the tumors; therefore, accurate modeling of lethally damaged cells is necessary for tumor-volume models and comparison to imaging data. The TCP is defined only by the number of viable clonogenic cells (proportional to n_1 and n_3) and the number of lethally damaged clonogenic cells (proportional to n_2 and n_4) can be omitted. However, tumor re-oxygenation is launched by the decay and removal of lethally damaged oxygenated cells. Therefore, the lethally damaged cells can indirectly affect the TCP because the rate at which they are removed from the tumor may affect re-oxygenation rate and, therefore, radiosensitivity of viable cells n_1 and n_3.

The 4-level model does not make any assumptions about tumor geometry; therefore, it can be applied to any tumor geometry. Derivation of the 4-level model does not make any assumptions about tumor geometry or location of cells; however, two spatially distinct regions that contain oxygenated and hypoxic cells are implicitly assumed. We also note that FMISO PET hypoxia images indicate that there is a threshold in the number of image pixels when the values of tumor/blood ratio = 1.2 [19]. This result indicates a spatially distinct hypoxia region; therefore, two discrete levels of hypoxia may be considered feasible. This two-level approach can also be generalized for any number of levels of the OER.

2.2.2 Cell Survival and Lethal Damage

The following assumptions are used to derive equations of the 4-level tumor model for fractionated radiotherapy with instantaneous irradiation with fractional dose d. Tumor irradiation causes death of a certain percentage of viable oxygenated and hypoxic cells. First, the cell survival curve $S(d)$ defines the relative number of cells which survive; therefore, the relative number of cells which are killed by radiation is given by $1 - S(d)$. Second, the irradiation with the fractional dose d does not affect the cells that are already lethally damaged; therefore, the new lethally damaged cells must be added to the previously lethally damaged cells in the tumor. Based on these assumptions, the following equations of the 4-level tumor model describe the cell numbers after instantaneous irradiation with the fractional dose d:

$$\overline{n}_{1,k} = n_{1,k} S_{ox}, \tag{2.2}$$

$$\overline{n}_{2,k} = n_{2,k} + n_{1,k}(1 - S_{ox}), \tag{2.3}$$

$$\overline{n}_{3,k} = n_{3,k} S_{hyp}, \tag{2.4}$$

$$\overline{n}_{4,k} = n_{4,k} + n_{3,k}\left(1 - S_{hyp}\right), \tag{2.5}$$

where $S_{ox} = S(d_{\mathrm{ox}})$ and $S_{hyp} = S(d_{\mathrm{hyp}}, \mathrm{OER})$ are the cell surviving fractions for oxygenated and hypoxic cells. The survival of oxygenated and hypoxic cells can be modeled using the LQ cell survival model given by Eq. 1.15: The parameters $\alpha_{hyp,}$ and β_{hyp} of the LQ model for hypoxic cells are presented as a function of parameters $\alpha_{ox,}$ β_{ox} for completely oxygenated cells and the OER. In the approximation of the 4-level tumor response model, we use two discrete levels of oxygen pressure with OER = 1 for well-oxygenated cells and for OER = $\mathrm{OER}_{\max}$ for hypoxic cells.

2.2.3 *Cell Disintegration, Repopulation and Oxygenation*

(a) **Cell Disintegration**

After instantaneous irradiation, the 4-levels of tumors cell population change due to removal of lethally damaged cells, reoxygenation, and repopulation. The most common form of cell death from radiation is mitotic death where a cell disintegrates at the first or subsequent division following irradiation [20]. The lethally damaged cells exist in some intact form and contribute to the tumor volume even though they are not able to proliferate. The disintegration of oxygenated lethally damaged cells can be modeled using the exponential decay with the decay constant $\mu = \ln2/T_{1/2}$ where $T_{1/2}$ is the half-life. The half-life $T_{1/2}$ can be evaluated using the variation with time of a large population of lethally damaged oxygenated cells in vivo. The large population of lethally oxygenated cells can be found, for instance, at the end of radiotherapy treatment when the tumor cells are assumed to be damaged, and tumor is reoxygenated. The half-life of lethally damaged cells has been evaluated using clinical data on the tumor-size variation after stereotactic radiotherapy [21, 22]. We note that exponential cell death after damage was also observed in other biological systems, for instance, inherited neuronal degenerations [23].

The disintegration model is different for lethally damaged oxygenated and hypoxic cells. The disintegration rate may be associated with the proliferation rate because both parameters are based on the cell cycle; therefore, the faster the cells proliferate the faster they disintegrate. The hypoxic cells have been effectively removed from the dividing population of cells in the body because they are resting in the G_0 phase. Hypoxic cells can remain in G_0 for days, weeks, or years, but can be stimulated to return to the cell-division cycle. It is well known that the hypoxic cells can remain in the tumor after death for a significant amount of time building necrotic mass. Therefore, we assume that the disintegration and removal of damaged hypoxic cells is much slower than the disintegration and removal of damaged oxygenated cells. The disintegration of lethally damaged hypoxic cells is dominated by reoxygenation rate; therefore, lethally damaged hypoxic cells disintegrate when they become reoxygenated.

(b) **Tumor Reoxygenation**

The time-dependent measurements of hypoxic volume during radiotherapy using FMISO PET demonstrated that the hypoxic volume changes for most patients with time following exponential decay with a small decay constant [24]. The time-dependent measurements of hypoxic volume and corresponding exponent functions are shown in Fig. 2.2 for two sample patients. Therefore, we utilize a constant fractional tumor re-oxygenation rate A that corresponds to exponential decay of hypoxic volume [16, 25]. The model can be modified to include time-dependent re-oxygenation rate $A(t)$ if needed.

(c) **Cell Proliferation and Final Equations**

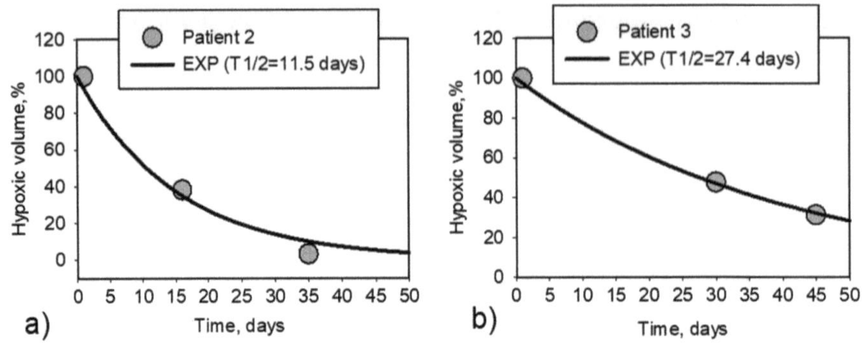

Fig. 2.2 Exponential approximation for the variation of hypoxic volume during radiotherapy for non-small lung cancer for two patients (**a**) and (**b**) measured using FMISO PET

We assume that repopulation of viable oxygenated cells is governed by exponential growth with the constant $\lambda = \ln 2/T_{pot}$ that is related to potential doubling time T_{pot} [1, 2]. Considering repopulation, removal of lethally damaged cells, and reoxygenation after irradiation with fractional dose d_k, the 4-levels of tumors cell population change after a time interval Δt based on the following equations describing relationship between dose fractions k and $k+1$ [16]:

$$n_{1,k+1} = \overline{n}_{1,k} \exp(\lambda_d \Delta t) + \overline{n}_{3,k}\mathrm{A}\,, \tag{2.6}$$

$$n_{2,k+1} = \overline{n}_{2,k} \exp\left(-\lambda_{1/2}\Delta t\right) + \overline{n}_{4,k}\mathrm{A}, \tag{2.6}$$

$$n_{3,k+1} = \overline{n}_{3,k}(1 - \mathrm{A}), \tag{2.8}$$

$$n_{4,k+1} = \overline{n}_{4,k}(1 - \mathrm{A}), \tag{2.9}$$

where A is the tumor-volume reoxygenation fraction; $\lambda_d = \ln 2/T_{pot}$ is the proliferation constant where T_{pot} is the potential doubling time of living cells; $\lambda_{1/2} = \ln 2/T_{1/2}$ is the disintegration constant in which $T_{1/2}$ is the half-life of lethally damaged cells.

2.2.4 Equation of 4-level Tumor Response Model

Combining Eqs. 2.2–2.5 for the instantaneous tumor irradiation and Eqs. 2.6–2.9 for cell disintegration, reoxygenation and proliferation, we obtain that the total variation of 4-levels of tumor cell population between the dose fractions is given by:

$$n_{1,k+1} = n_{1,k} S_{ox} e^{\lambda_d \Delta t} + n_{3,k} S_{hyp}\mathrm{A}, \tag{2.10}$$

$$n_{2,k+1} = \left(n_{2,k} + n_{1,k}(1 - S_{ox})\right)e^{-\lambda_{1/2}\Delta t} + A\left(n_{4,k} + n_{3,k}\left(1 - S_{hyp}\right)\right) \quad (2.11)$$

$$n_{3,k+1} = n_{3,k}S_{hyp}(1 - \mathrm{A}), \quad (2.12)$$

$$n_{4,k+1} = \left(n_{4,k} + n_{3,k}\left(1 - S_{hyp}\right)\right)(1 - \mathrm{A}). \quad (2.13)$$

Following Eqs. 2.10–2.13, the total number of cells in tumor N during radiation therapy changes due to proliferation of dividing viable cells and disintegration of lethally damaged cells. Cell killing and re-oxygenation generate cell transition between cellular levels and do not change the total cell number N in the system. An example of variation of four tumor cell levels with time during fractionated radiotherapy is shown in Fig. 2.3. The 4-level tumor model do not make any assumptions about radiotherapy or chemotherapy killing models because only cell surviving fractions are used. Therefore, these equations can be applied to radiotherapy alone, or concomitant radiotherapy and chemotherapy treatments. This conclusion is supported by successful fitting to measured tumor-volume curves as is shown below in this Chapter.

The system of Eqs. 2.10–2.13 describing the 4-level tumor response model is complex with four parameters $n_{i,0}, i = \overline{1,4}$ to describe the initial tumor condition and five free parameters ($S_{\mathrm{ox}}, S_{\mathrm{hyp}}, \lambda_d, \lambda_{1/2}, \mathrm{A}$) to simulate the radiotherapy response. This system of equations can be used for direct simulation of radiotherapy responses, assuming all initial conditions and parameters are known [16]. However, this system of equations can become overdetermined for solving the inverse problems when the

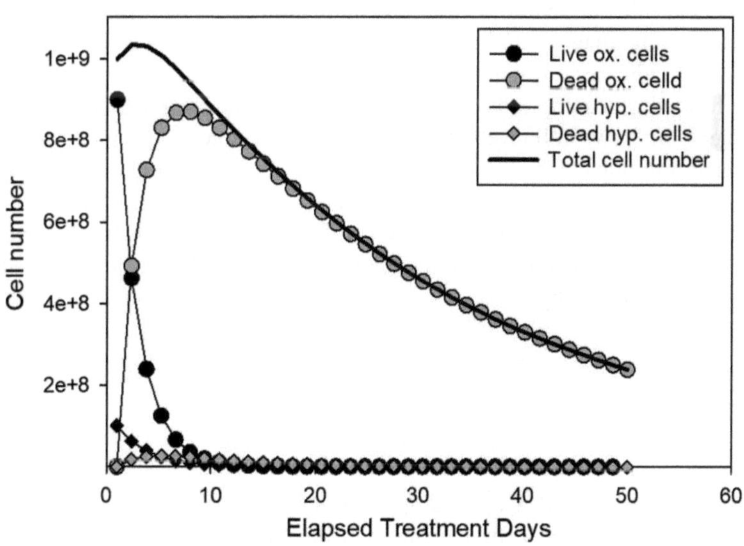

Fig. 2.3 Variation of 4-level tumor cell population during fractionated radiotherapy [16]

reconstruction of model parameters from imaging or measured data are needed [7, 14, 15]. In this situation the problem oriented simplified versions of the 4-level tumor response model should be used with additional constraints on model parameters if needed. The 2-level cell population tumor response models considered below can be considered as an example of these problem-oriented tumor response models.

2.3 Two-Level Tumor Response Models

2.3.1 Tumor Response Model for TCP Assessment

The analysis of the TCP in fractionated radiotherapy requires application of tumor response models that can describe major components of tumor response to irradiation such as intrinsic radiosensitivity, proliferation, and hypoxia but use as few parameters as possible for robust comparison with clinical data. It is important that these parameters could be related to the measurements with predictive assays such as the parameters of the LQ model and cellular proliferation constant. Some parameters such as the tumor hypoxic fraction and re-oxygenation rate can be evaluated from functional imaging. As we already mentioned, the lethally damaged cells have no direct impact on the TCP and can be omitted in this model. Therefore, we can employ the 4-level tumor response model without lethally damaged cells that results in a 2-level tumor response model. The different versions of 2-level tumor response model have been used to address the problems associated with tumor response to radiation therapy and TCP analysis [25–27]. This model also allows us to obtain analytical solutions and other advantages of this model include possible evaluation of model parameters using imaging studies and predictive assays and as well as application to concomitant radiotherapy and chemotherapy.

The following assumptions can be used for the tumor response model for TCP assessment. Only two cell levels are considered: the viable oxygenated and hypoxic clonogenic cells that will be denoted as n_{ox} and n_{hyp} correspondingly. During fractionated radiotherapy treatment, a tumor is irradiated with K fractional doses separated by a time interval Δt. Considering the models of exponential cellular proliferation and re-oxygenation, the population of tumor clonogenic cells will change after irradiation at each dose fraction k by

$$n_{ox,k+1} = n_{ox,k} S_{ox} \exp(\lambda_d \Delta t_k) + S_{hyp} n_{hyp,k} A_k, \tag{2.14}$$

$$n_{hyp,k+1} = n_{hyp,k} S_{hyp} (1 - A_k), \tag{2.15}$$

where $S_{ox} = S(d_{\mathrm{ox}})$ and $S_{hyp} = S(d_{\mathrm{hyp}}, \mathrm{OER})$ are the cell survival fractions of oxygenated and hypoxic cells calculated using the LQ model; $\lambda_d = \ln 2/T_{\mathrm{pot}}$ is the proliferation constant where T_{pot} is the potential doubling time, and A_k is the reoxygenation rate that describes a fraction of the hypoxic volume that becomes reoxygenated after irradiation with the fractional dose. Cellular proliferation and re-oxygenation are time dependent processes; therefore, the weekends should be included to correctly describe these processes. To model the weekends, proliferation time can be increased by the factor of 3 and daily re-oxygenation rate was applied 3 times.

Reoxygenation in the 2-level tumor response model is described using only two parameters: initial hypoxic fraction and reoxygenation rate. Tumor cells in the hypoxic region can change their status due to re-oxygenation. Reoxygenation rate is a measure that describes the proportion of the hypoxic fraction that becomes re-oxygenated after each dose fraction. These two parameters can be evaluated using serial imaging studies of hypoxia [24]. We note that other parameters of the model such as cell proliferation rate and cell survival fractions can be evaluated using serial imaging as well [6, 7]. Although more detailed models that describe several levels of cells that are related to hypoxia have been previously reported, we did not apply them in our study because they had more variables [9–13].

2.3.2 Tumor Response Model for Volume Assessment

Equations 2.10–2.13 of the 4-level tumor model describe a general case of a tumor with heterogeneous sensitivity that undergoes radiotherapy treatment. In some specific situations, the 4-level tumor model can be simplified to the 2-level tumor model for tumor volume assessment that is more robust for parameter optimization and data fitting. Some tumors in radiotherapy treatment can be completely oxygenated or quickly re-oxygenated within the first few fractions; therefore, the hypoxia parameters may not be important for these tumors. We also note that at the dose levels used for cancer radiotherapy, the number of viable clonogenic cells falls quickly during the first few fractions to a level where they do not affect the tumor volume dynamics. Most clonogenic cells are killed by radiation within the first few fractions; hence, the cell killing, and cell-growth parameters may be not important for tumor-volume dynamic during most of treatment time.

The 2-level tumor response model for tumor volume assessment divides the entire tumor cell population into two subpopulations of viable cells and lethally damaged cells and considers the mechanisms of cell killing, cell proliferation, and cell disintegration. The effects of tumor heterogeneity, due specifically to tumor hypoxia, are considered in the two-level cell population model by weighing the kinetics of oxygenated and hypoxic cells in the effective model parameters.

We note that one of the main approximations of the 2-level tumor model is the homogenization of heterogeneous tumor response. This approach may be valid because the hypoxic fraction is small, for example, for a large fraction of head and neck cancer patients as measured by PET and polarographic oxygen electrodes [19, 28, 29], and, additionally, the re-oxygenation process may be rapid [24]. The 2-level tumor model is attractive because of the transparency of its algorithm; however, this model is a fundamental algorithm, and more complicated models can be obtained from it as corrections.

The following equations for the 2-level tumor model can be obtained from the 4-level tumor model assuming that hypoxic tumor fraction is equal to zero. The equations describe the variation in the numbers of viable cells n_1 and lethally damaged cells n_2 between dose fractions k and $k + 1$ after irradiation with the fractional dose (d_k) and after a time interval Δt [6, 7]:

$$n_{1,k+1} = SF(d)n_{1,k} \exp(\lambda_d \Delta t), \tag{2.16}$$

$$n_{2,k+1} = \left[n_{2,k} + n_{1,k}(1 - SF(d))\right] \exp\left(-\lambda_{1/2}\Delta t\right). \tag{2.17}$$

These equations include the following parameters: *SF(d)* is the effective survival fraction of viable cells; $\lambda_{\mathrm{d}} = \ln 2/T_{\mathrm{d}}$ is the proliferation constant, where T_{d} is the effective doubling time of living cells; $\lambda_{1/2} = \ln 2/T_{1/2}$ is the clearance constant, where $T_{1/2}$ is the effective half-life of lethally damaged cells. The half-life $T_{1/2}$ of lethally damaged cells is defined as the time it takes for the number of cells to decay to half the initial value. Equations 2.16 and 2.17 are obtained by assuming the probability that a cell will not survive is 1- $SF(d)$ if $SF(d)$ is the probability to survive. Further, an exponential clearance with the half-life $T_{1/2}$ is assumed for lethally damaged cells, which is not altered by subsequent irradiation with fractional doses. The effects of tumor heterogeneity, for example, due to hypoxia are considered in the 2-level tumor model implicitly by weighing the effective model parameters for the entire tumor volume.

2.4 Half-Life of Lethally Damaged Cells

2.4.1 Disintegration of Lethally Damaged Cells

A model for disintegration of lethally damaged cells is a key component in modelling of tumor response to radiotherapy especially for comparison with imaging data. These cells are damaged by radiation with the probability described, for instance, by the LQ-model; however, they exist in some intact form until disintegration and contribute to the tumor volume even though they are not able to proliferate. It is usually assumed that a cell damaged by radiation at the dose levels used in radiotherapy disintegrates at the first or a subsequent division after radiation damage. This mechanism called mitotic death is a common form of cell death after irradiation; however, the probability of other disintegration mechanisms like apoptosis can be also included in our model [20]. We show that the time-dependent disintegration of tumor cells which do not survive radiotherapy can be described using a simple analytical function. We assume that the cell disintegration rate is proportional to the number of intact damaged cells; therefore, we can write the following differential equation for the time-dependent number of lethally damaged cells $n(t)$:

$$\frac{dn(t)}{dt} = -\lambda_{1/2}(t)n(t), \tag{2.18}$$

where $\lambda_{1/2}(t)$ is the cell disintegration constant.

We further introduce an approximation of constant disintegration risk that assumes the disintegration constant is time independent $\lambda_{1/2}(t) = \lambda_{1/2}$. Equation 2.18 with $\lambda_{1/2}(t) = \lambda_{1/2}$ has an analytical exponential decay solution which is given by

$$n(t) = n(t_0)\exp\left(-\lambda_{1/2}t\right), \tag{2.19}$$

where $n(t_0)$ is the number of dead intact cells at the initial time t_0. This approach is like the mathematical formalism used to describe the time-dependent kinetics of cell death in inherited neuronal degenerations [23].

Similar to modeling of radioactive decay, we introduce a parameter called half-life for the biological decay of the population of damaged cells. The half-life is defined as the time required for the number of damaged cells to decay to half of their initial value. The half-life of the damaged cell population is related to the decay constant $\lambda_{1/2}$ as

$$T_{1/2} = \frac{\ln 2}{\lambda_{1.2}}. \tag{2.20}$$

The key problem for practical applications of this approach is to evaluate the half-life of the population of damaged cells. We have already mentioned that damaged cells disintegrate at the first attempted division; therefore, the disintegration rate may be associated with the proliferation rate because both parameters are related to the cell cycle. It is convenient to assume that the half-life $T_{1/2}$ is linearly related to the potential doubling time T_{pot} as

$$T_{1/2} = bT_{pot}, b > ln2. \tag{2.21}$$

If we consider cell disintegration at the first or subsequent division, we can further establish that $T_{1/2} > \ln 2T_{\text{pot}}$ that allows for the parameter $b > \ln 2$. We utilized here the relationship $T_{1/2} > \ln 2T_{\text{a}}$ between the half-life $T_{1/2}$ and the mean life T_{a}. Obviously, we obtain $T_{\text{a}} = T_{\text{pot}}$ if all damaged cells disintegrate at the first division. For detailed evaluation of half-life $T_{1/2}$, we must study the variation with time of a large population of damaged cells in-vivo. The large population of damaged cells can be found, for instance, at the end of radiotherapy treatment where the entire population of cancer cells is assumed to have not survived irradiation.

2.4.2 Timor-Volume Variation After Radiotherapy

It is difficult to derive the cell disintegration constants using the data on tumor-volume variation during radiotherapy because the tumor volume is defined by the kinetics of proliferating and damaged cells. However, we can assume that the entire cell population did not survive the irradiation after radiotherapy; therefore. only one cell subpopulation of lethally damaged cells should be considered.

One of the possible ways to study the kinetics of cell disintegration is to evaluate the tumor volume after radiotherapy because we can assume that the entire tumor cell population did not survive radiotherapy. Clinical data on tumor-volume variation after radiotherapy are available, for instance, for solitary lung tumors treated with SBRT [21]. This is a relatively simple mathematical problem because we must evaluate only one cell subpopulation damaged by radiation and unable to proliferate.

Assessment of radiobiological parameters of tumor cells using tumor models and measured tumor volume variation curves requires an assumption of linear relationship between the tumor volume and the total number of tumor cells during radiotherapy treatment. To evaluate the kinetics of cell disintegration based on the volumetric tumor measurements, we must assume a linear relationship between the tumor volume $V(t)$ and the cell number $n(t)$:

$$V(t) = \nu n(t), \tag{2.22}$$

where ν is a constant which includes the cell volume and the volume of the related intercellular space. This assumption is supported by radiobiological experiments which show that the cell concentration in tumors may decrease after irradiation, however, it quickly returns to the pre-irradiation level for the fractional doses used in conventional fractionation.

Experiments with animal irradiation indicate that the mean cell concentration in tumors can be a function of delivered radiation dose and time. Therefore, the gross changes in the tumor volume after irradiation are not always a reliable indicator of microscopic changes in the cell number. However, the experimental data indicates that the mean cell concentration returns to the near-normal values after a limited time between 1 and 7 days [30]. For instance, a 25% reduction in the mean cell concentration was observed after irradiation with 6 Gy dose and 1 day of recovery time was needed. After irradiation with a larger dose of 30 Gy, a 50% reduction in the mean cell concentration and 7 days of recovery time was measured. Therefore, we believe that the linear relationship between the cell number and the tumor volume is a reasonable approximation for tumor volume simulation in the clinical study analyzed in this article. In this clinical study, tumors have been monitored for several months after radiotherapy.

To validate the exponential model and determine the decay parameters, we use clinical data on tumor size variation after SBRT [21]. This clinical study includes analysis of the shrinkage of solitary lung tumors in 31 patients after administering a total dose of 48 Gy in 4 fractions using conformal SBRT. Considering that the cell survival fraction given by Eq. 1.13, we can assume that the entire cell population tumors did not survive the radiotherapy at this dose; therefore, the tumor volume variation after radiotherapy should be defined by the disintegration and removal of damaged cells which are not able to proliferate. This is probably true for most patients because tumor control was not obtained in only two cases.

2.4.3 Half-Life Derived from Tumor Volume Variation

The full clinical study on tumor size variation after SBRT [21] includes 15 adenocarcinoma cases, 9 SCC cases, 4 metastasis cases, 2 unknown cases and 1 small cell carcinoma case. We have separated the adenocarcinoma and SCC cases and excluded all other cases. The tumor shrinkage was observed in all adenocarcinoma and SCC patients; however, not all of them had enough tumor size measurements to perform a reasonable fitting. One of selection criteria was the availability of at least three tumor measurements for each patient and at least one tumor measurement within the first four months after treatment. Therefore, the patients for the fitting have been selected based on the number and density of tumor measurements available for each case. To determine the half-life $T_{1/2}$- of the exponential cell disintegration, we perform a computerized fitting of exponent function to selected clinical data for 7 cases of adenocarcinoma and 7 cases of squamous cell carcinoma (SCC).

The tumor size variation with time in the clinical study was determined by taking CT images after radiotherapy. We note that the tumor size has been measured using the largest transversal cross section A of the tumor. To evaluate the tumor volume variation, we computed the relative change of value $AA^{1/2}$ which presents the relative volumetric change under the assumption of uniform tumor shrinkage. The relative change in the value $AA^{1/2}$ is shown in Fig. 2.4 for one SCC case and one adenocarcinoma case. One can see that there is a small constant component in the volume variation does not change with time; therefore, this component may be due to residual irregular density as radiation fibrosis and was, therefore, subtracted from volume variation in our computations.

The mathematical function used for fitting is presented as a weighted linear combination of an exponent function and a constant:

$$f(t) = (1 - w)\exp\left(\lambda_{1/2}t\right) + w, \tag{2.23}$$

where w is a constant value between 0.0 and 1.0. The constant value w is included to subtract the constant saturation value in the clinical data during the optimization process. The fitting of the exponential function is done using the least squares minimization based on a version of a quasi-Newton method called the L-BFGB-B algorithm [31]. Sample results of fitting are shown in Fig. 2.4a for one SCC patient and Fig. 2.4b for one adenocarcinoma patient. We note that the approximation of tumor-volume regression curves by the exponential decay functions supports the linear relationship between the tumor volume and cell number accepted in Eq. 2.22.

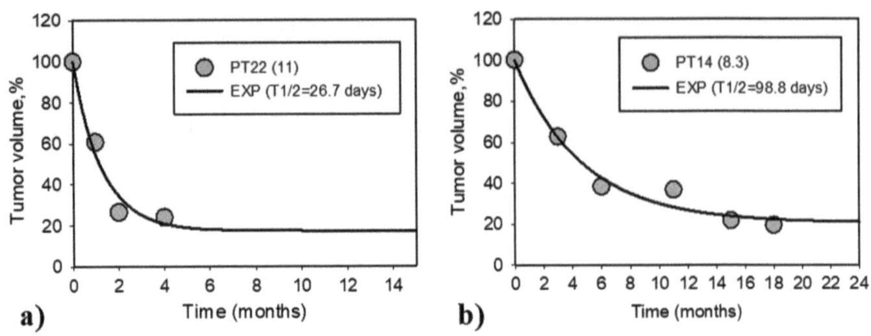

Fig. 2.4 Measured and computed tumor volume variation after SBRT for non-small lung cancer during radiotherapy for two patients PT22 (SCC) and PT14 (adenocarcinoma). Numbers in parentheses show tumor volumes for each patient in cubic centimeters

The linear relationship is one of the mathematical conditions of the exponential tumor-volume variation.

We obtained the average half-life $T_{1/2} = 28.2$ days and standard deviation $\sigma = 8.6$ days for SCC and the average half-life $T_{1/2} = 72.4$ days and standard deviation $\sigma = 22.8$ days for adenocarcinoma. Assuming the potential doubling time $T_{\text{pot}} = 7.7$ days for adenocarcinoma and $T_{\text{pot}} = 8.5$ days for SCC [32], we obtain the average disintegration parameter $\text{b} = 9.9$ for adenocarcinoma and $\text{b} = 3.3$ for SCC. These data indicate that that the cell cycling after lethal radiation damage before disintegration in adenocarcinoma is approximately three times longer than in SCC. Using the half-life $T_{1/2}$, we can also evaluate the mean-life of a single cell after radiation damage. The mean life is given by $T_{\text{a}} = T_{1/2}/\ln 2$; therefore, we obtain the average mean life of $T_{\text{a}} = 40.7$ days for SCC and $T_{\text{a}} = 105.5$ days for adenocarcinoma.

The standard deviation of half-life for both adenocarcinoma and SCC is approximately 30% of the average half-life. This relatively large standard deviation can be explained by several reasons: (1) patient heterogeneity is usually a reason of large standard deviation of radiobiological parameters [33]; (2) small number of cases in evaluation and (3) lung tumors may not be completely re-oxygenated at the end of SBRT because this treatment is relatively short; therefore, lethally damaged oxygenated and hypoxic cells may be present in tumors. Hypoxic cells are practically removed from the dividing population of cells in the tumor because they are resting in the G_0 phase. Hypoxic cells can remain in G_0 phase for days, weeks, or years, but can be stimulated to return to the cell division cycle. Therefore, the disintegration of lethally hypoxic cells may depend on tumor re-oxygenation rate and the residual tumor hypoxia may affect heterogeneity of measured half-life value. According to the data obtained using FMISO PET, hypoxia for non-small lung cancer can be relatively large with the median value of 47.6% and the maximum value of 94.7% [19].

2.5 Parameter Estimation in Tumor Response Models

2.5.1 Ill-Posed Problem in Reconstruction of Model Parameter

The parameters of the 2-level and 4-level cell population tumor models can be determined by computerized fitting to the patient-specific tumor-volume variation curves. A simulated annealing optimization algorithm and a least squared objective function can be used for patient-specific optimization of the tumor models. The derived estimates of patient-specific parameters related to cell survival, proliferation and cell clearance can be used to calculate the patient-specific TCP.

However, the reconstruction of radiobiological parameters from the serial tumor imaging data using the kinetic models of tumor response is ill-posed and may produce large nonphysical fluctuations in the reconstructed radiobiological parameters. The tumor response models involve a sum of several exponentials describing cell proliferation, cell killing, and cell clearance. Therefore, the total cell number N(t) can be formally written in the following form:

$$N(t) = \sum_{l=1}^{L} C_l \exp(-\lambda_l t), \tag{2.24}$$

that represents a discretized form of the Fredholm integral equation of the first kind [14, 15]:

$$N(t) = \int_{0}^{\infty} C(\lambda) \exp(-\lambda t) d\lambda \tag{2.25}$$

The Fredholm integral equation of the first kind is ill-posed in the sense that small errors in the input measured data $N(t)$ can produce large errors in the function $C(\lambda)$; therefore, the problem of reconstruction of radiobiological parameters from imaging data becomes an ill-posed inverse problem. As will be shown below, a direct application of Eq. 2.25 produces large nonphysical fluctuations in the reconstructed parameters. Special stabilization algorithms should be applied to reduce the errors in the reconstructed parameters.

The ill-posedness can be overcome by using constraints on the model parameters. For example, Eq. 2.24 becomes a well-posed linear problem if the parameters λ_l, $l = \overline{1, L}$ are known. For example, using a constant value of $\lambda_{\mathrm{d}} = \lambda_{mean}$ in Eq. 2.16–2.17 where λ_{mean} is the mean value of the expected proliferation constant, we can evaluate the cell surviving fractions [6]. However, the constraint on cell doubling time may introduce additional errors because the cell surviving fractions reconstructed with a mean value of proliferation constant may include both inter-patient variation in tumor radiosensitivity and proliferation constant.

2.5.2 Least-Squared Minimization with Regularization

A more advanced way to stabilize numerical solutions and reduce errors is to use the variational regularization method that usually is referred to in the literature as Tikhonov regularization [15]. The Tikhonov regularization is obtained by adding to the least squares objective function to find the vector $\boldsymbol{C_\alpha} = \{C_l, l = \overline{1, L}\}$:

$$\boldsymbol{C_\alpha} = argmin\left[\sum_{k=1}^{K}\left(V_k(\boldsymbol{C_\alpha}) - V_k^M\right)^2 + \alpha\Omega_M(\boldsymbol{C_\alpha})\right] \quad (2.26)$$

a special smoothing term that can be written in a general form as:

$$\Omega_M[C(\xi] = \sum_{m=0}^{M}\int_a^b d\xi p_m(\xi)\left[\frac{d^m C(\xi)}{d\xi^m}\right]^2. \quad (2.27)$$

In Eq. 2.26, V_k and V_k^M are the modelled and measured tumor volume at the time interval k, respectively, and α—is the regularization parameter. For example, the simplest zeroth order form of the smoothing term is given by:

$$\Omega_0[\boldsymbol{C_\alpha}] = \sum_{l=0}^{L} C_l^2 \quad (2.28)$$

For instance, the following zeroth order regularization term can be used for the 2-level cell population tumor model:

$$\Omega_0[SF_d, \lambda_d] = SF_d^2 + \lambda_d^2. \quad (2.29)$$

The key problem in the variational regularization methods is the evaluation of an optimal value of the regularization parameter α—a weighting coefficient for the stabilizing functional. To select an optimal value of regularization parameter, we apply the L-curve method which was developed to determine the regularization parameter in discrete ill-posed problems [34]. The L-curve method is based on finding the regularization parameter that minimizes the residual norm, which is a measure of accuracy of fit, and the solution norm, which is a measure of smoothness of solution. The L-curve method consists in the application of the following variables $\xi(\alpha)$ and $\eta(\alpha)$:

$$\xi(\alpha) = \left[\sum_{k=1}^{K}\left(V_k(\boldsymbol{C_\alpha}) - V_k^M\right)^2\right]^{1/2}, \quad (2.30)$$

$$\eta(\alpha) = [\Omega_M(\boldsymbol{C_\alpha})]^{1/2}. \quad (2.31)$$

The first variable $\xi(\alpha)$ describes the accuracy of fit of the model-based tumor volume to the measured imaging-based volume. The second variable $\eta(\alpha)$ relates to the stabilizing functional Ω_M and presents a norm of smoothness of solution. The optimal value of the regularization parameter α is one that minimizes both variables $\xi(\alpha)$ and $\eta(\alpha)$. The main idea of the L-method is to plot the smoothing norm $\eta(\alpha)$ as a function of the residual norm $\xi(\alpha)$ for all values of the regularization parameter α. It was previously shown that this characteristic function has an L-shaped form with distinctive vertical and horizontal parts. The optimal value of the regularization parameter can be found in the vicinity of the "corner" of the L-curve where both values of smoothing norm $\eta(\alpha)$ and residual norm $\xi(\alpha)$ achieve their minimum.

2.6 Parameter Estimation Using Clinical Data

2.6.1 Radiotherapy Problem with Serial CT Imaging

Testing stability of reconstruction of radiobiological parameters using the 2-level cell population model requires a special stability test. The parameter reconstruction for each patient involves only three variables and, therefore, it is difficult to see and study the nonphysical fluctuations using only three variables. This problem is different from the reconstruction of energy spectra or beam intensity profiles for intensity-modulated radiation therapy (IMRT), where a significant number of energy or spatial bins (100–200 or more) is available to monitor and separate the nonphysical fluctuations [35, 36]. To overcome this problem, we can reconstruct radiobiological parameters for a patient group in which we could expect that these radiobiological parameters would be distributed according to a normal distribution. A deviation from the expected normal distribution can be considered as instability of the reconstruction method.

To study the stability of the reconstruction method, we use the tumor-volume variation data from clinical study for 14 head and neck cancer patients where an integrated CT/ linear particle accelerator (linac) system was used for radiotherapy treatment and tumor volume measurements [37]. Patients in this data set were treated with 1.8 Gy or 2 Gy dose fraction and almost all patients received concurrent chemotherapy. Only gross tumor volumes (GTVs) are simulated in this article. The CT images for one patient are shown in Fig. 2.5. The GTV variation curves have been normalized to the initial tumor volume at the start of radiation treatment. Accurate normalization is an important factor in comparing the simulations and measured data because the initial volume value affects all measured tumor-volume points.

This clinical study for head and neck cancer has several advantages, compared to other published data on tumor-volume variation during radiotherapy treatment, that are important for validation of the model. These advantages include minimized uncertainties in the tumor-volume definition associated with high contrast and resolution of the fan-beam CT scanner, small uncertainties associated with breathing

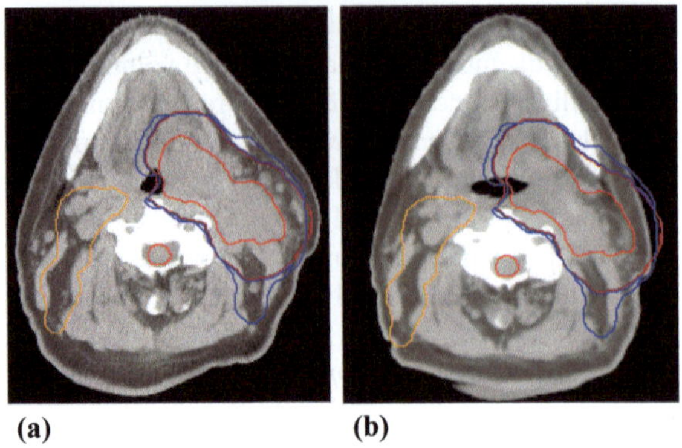

Fig. 2.5 Shrinkage of head-and-neck tumor during radiotherapy treatment: **a** planning CT; **b** CT taken during radiotherapy treatment

motion, and a significant number of tumor-volume points (CT images were taken three times a week). It is also important that previous PET studies with hypoxia tracers have shown that the pretreatment hypoxic volumes were relatively small for head and neck cancer (median value 8.8%) [19] that makes the 2-level model an acceptable approximation. We note that pretreatment hypoxic volumes for the head and neck cancer measured using polarographic oxygen electrodes were somewhat larger, for example, 15% median value of hypoxic fraction was reported in [29] and 43% of 23 patients having hypoxic fraction larger than 20% was measured in [28]. However, the hypoxic fraction is relatively small for more than half of all patients even for the electrode data; therefore, the 2-level model could be valid for many patients in a study group.

2.6.2 Reconstruction Algorithm

Assuming the linear dependence between the total cell number and tumor volume, the tumor volume V_{k+1} after irradiation with dose fraction k is calculated using the following equation:

$$V_{k+1} = \nu\left(n_{1,k+1} + n_{2,k+1}\right), \tag{2.28}$$

where $n_{1,k+1}$ and $n_{2,k+1}$ are the number of viable and lethally damaged cells and ν is a proportionality constant.

Data fitting is performed using a least squares objective function with the simulated annealing optimization algorithm for minimum search. A simulated annealing optimization algorithm and the least squared objective function are used for the patient-specific optimization of the model. We consider the simulated annealing algorithm the best choice for this optimization problem because this method allows the global minimum of a complex objective function to be found without evaluating its gradients. To find the minimum of the least squared objective function, we can use the implementation of simulated annealing [38] that is based on the continuous simulated annealing global optimization algorithm [39]. The reconstruction was performed without stabilization algorithms and with a stabilization algorithm based on variational regularization.

2.6.3 Intrinsic Radiosensitivity and Proliferation Constant

Intrinsic tumor radiosensitivity characterized by cell surviving fractions is a key parameter in the evaluation of tumor response to radiation therapy and markedly contributes to inter-patient variation of tumor control probability. Practically, intrinsic tumor radiosensitivity can be evaluated in a radiation therapy clinic only by using predictive assays. We show how the capabilities of serial imaging can be extended to evaluate intrinsic tumor radiosensitivity and proliferative potential using tumor response models. This method could serve as an alternative to predictive assays based on tumor biopsies in the case of fast proliferating tumors such as lung cancer, head and neck cancer, or cervical cancer.

In Fig. 2.6a–n, we display a comparison of the GTV variation obtained from the CT imaging data during radiation therapy and computed with the 2-level model for 14 patients. The original numbers are preserved for each patient. The case of variational regularization is shown; however, the fit to the clinical data is almost identical for the cases with and without regularization. Regularization mostly improves the reconstructed model parameters as shown below. The model describes trends in relative volumetric tumor changes observed in clinical data within the range of measurement errors. In particular, the model describes the initial "shoulder" part of the curves and the gradient of tumor-volume variation during treatment that is larger at the beginning and smaller toward the end of the treatment.

In Fig. 2.7, we show a comparison of probability density functions (PDFs) for cell surviving fractions SF_2 and cell number doubling time T_{d} in the study with the PDFs obtained by using in vitro predictive assays [33]. This comparison demonstrates that these PDFs do not agree. The reconstructed cell surviving fractions SF_2 and cell number doubling times T_d were obtained without a stabilization algorithm. The cell surviving fractions reconstructed for dose fractionation schedules with 1.8 Gy per fraction have been recalculated to 2.0 Gy per fraction using a simple exponential model of dose response. One can see that the SF_2 and T_d values reconstructed without stabilization algorithms exhibit significant nonphysical fluctuation between patients, sometimes reaching unfeasible values: for example, very small and large values of cell number doubling time.

To apply the variational regularization, the L-curve was computed for each patient and the optimal value of the regularization parameter was determined that minimized the residual and smoothing norms as described in Eqs. 2.30 and 2.31. An example of the L-curve for patient HN14 is shown in Fig. 2.8. This patient has the most remarkable L-shape of the L-curve. The regularization parameter was selected on the vertical part of the L-curve before entering the curvature part of the corner. The

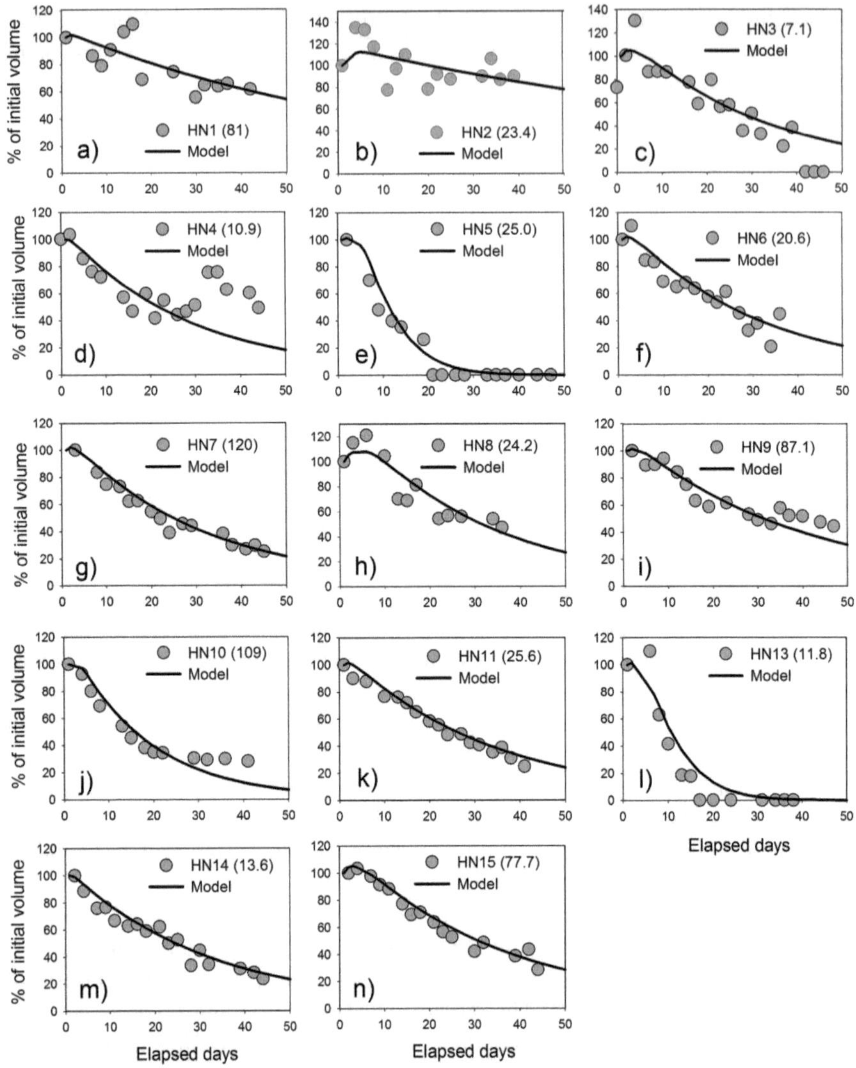

Fig. 2.6 Measured tumor-volume variation during radiotherapy for head and neck cancer patients HN01-HN15, and simulation by the two-level population tumor model. Numbers in parentheses show tumor volumes for each patient in cubic centimeters [7]

results of reconstruction of cell surviving fractions SF_2 and doubling time T_d with the variational regularization are shown in Fig. 2.9. The reconstructed values of SF_2 and T_{d} do not have nonphysical oscillations, and the corresponding PDFs are in good agreement with PDFs obtained from predictive assays, as can be seen in Fig. 2.9.

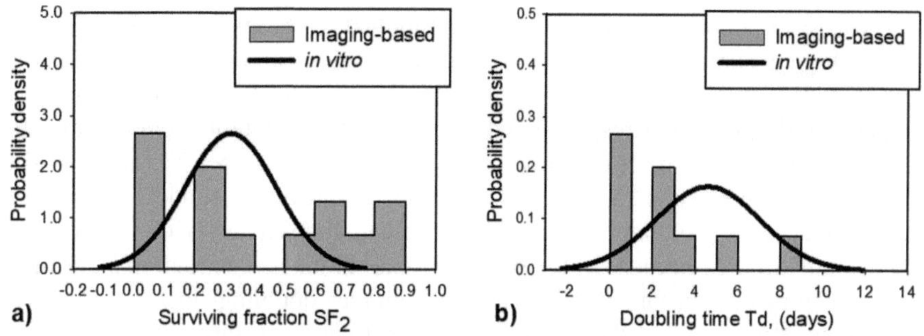

Fig. 2.7 Comparison of probability density functions for cell surviving fractions (**a**) and cell doubling time (**b**) reconstructed from imaging data without regularization and measured in vitro using predictive assays [7]

Fig. 2.8 An example of an L-curve for patient HN14

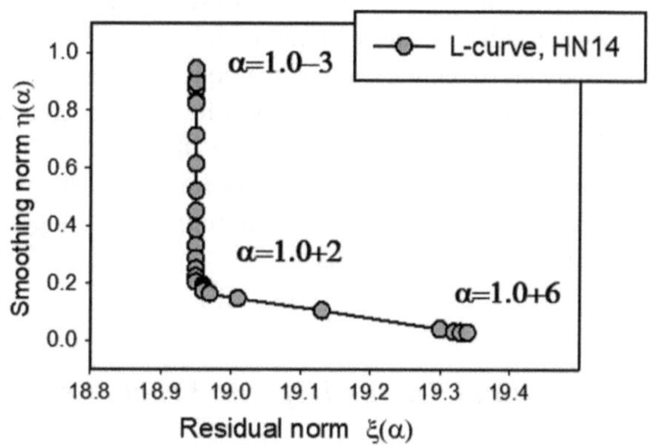

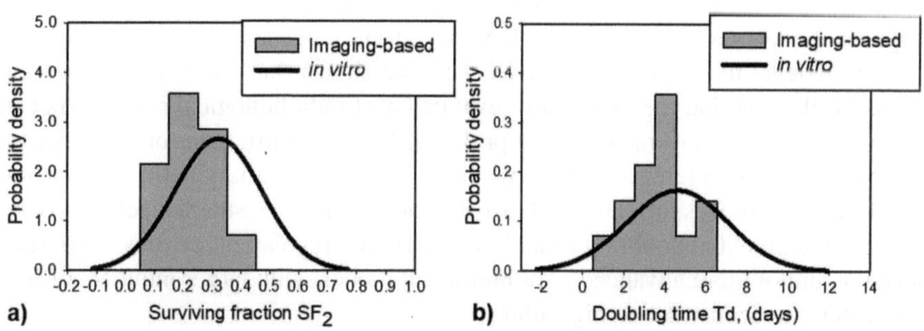

Fig. 2.9 Comparison of probability density functions (PDFs) for cell surviving fractions (**a**) and cell doubling time (**b**) reconstructed from imaging data with variational regularization and measured in vitro using predictive assays [7]

The 2-level population tumor model was applied to reconstruct patient-specific radiobiological parameters from tumor-volume variation curves measured using on-board CT imaging for head and neck cancer. The 2-level model is, probably, one of the simplest models of tumor response and has only three radiobiological parameters: cell surviving fraction SF_2, doubling time of viable cell number T_d, and half-life of lethally damaged cells $T_{1/2}$. In previous applications of this model, a constraint was introduced for the doubling time T_d to evaluate the cell surviving fraction SF_2 and the half-life $T_{1/2}$ because of the stability issues in the reconstruction algorithms [6]. An estimated constant value of the cell number doubling time T_d was used in the reconstruction, and the problem of reconstruction of three parameters was reduced to the reconstruction of two parameters. Using only an estimated value of T_d without allowing inter-patient variation inevitably introduces related errors in the reconstructed values of cell surviving fractions SF_2.

The patient-specific tumor response parameters have been determined by fitting the 2-level cell population tumor model to the GTV variation curves [7]. By applying a more advanced regularization technique, we were able to reconstruct not only the cell surviving fractions but also cell proliferation constants, considering the inter-patient variation for both parameters. Our results indicate that by using advanced stabilization methods, the number of free parameters in the model can be increased. It opens the way to the development of more advanced algorithms, which can reconstruct larger numbers of parameters.

2.6.4 Initial Tumor Hypoxic Fractions

The evaluation of radiobiological parameters using the serial imaging data and tumor response models shows that this problem may produce unstable solutions. Using the simplest 2-level tumor response model we have shown that these instabilities are related to the fact that the problem of evaluation of radiobiological parameters can be considered as an inverse ill-posed problem. Using in vitro data for radiobiological parameters as reference values, we have shown that variational regularization significantly improves the statistical distributions of the reconstructed cell surviving fraction and potential doubling time. Using variation regularization may encourage development of more advanced radiobiological models with larger numbers of free parameters in reconstruction algorithms.

Despite the implementation of advanced regularization techniques, fitting the 4-level response model with hypoxic cellular levels to measured data remains a challenging problem because of the significant number of radiobiological parameters and the uncertainties of measured data. We attempt to use the 4-level model with an additional constraint on re-oxygenation constant. We assume the re-oxygenation is slow and re-oxygenation constant $A = 0$ in Eqs. 2.10–2.13. This reduces the number of free parameters and allows one to estimate the initial tumor hypoxic fraction using serial volumetric imaging.

The fitting of the 4-level response model to the serial CT volumetric imaging data for head and neck cancer was performed using a least squares objective function and simulated annealing algorithm is shown in Fig. 2.10. The problem of reconstruction of radiobiological parameters from serial imaging data was considered as an inverse ill-posed problem described by the Fredholm integral equation of the first kind. The variational regularization was used to stabilize solutions. In Fig. 2.10, we show the simulation of tumor volume variation computed using the 2-level and 4-level tumor response models. Comparing Figs. 2.6 and 2.10, we can conclude that the 2-level response model can be sufficient in many cases; however, the 4-level response model can improve approximation for some patients. In Fig. 2.11, the distribution of reconstructed initial tumor volumetric hypoxic fractions (sum of n_3 and n_4) is compared with the distribution of hypoxic fractions measured using polarographic oxygen electrodes [28]. These two distributions are in good agreement.

The agreement of measured and reconstructed from imaging data tumor hypoxic fractions supports the validity of the 4-level response model. These results support the hypothesis that the hypoxic tumor fraction can be computed using radiobiological

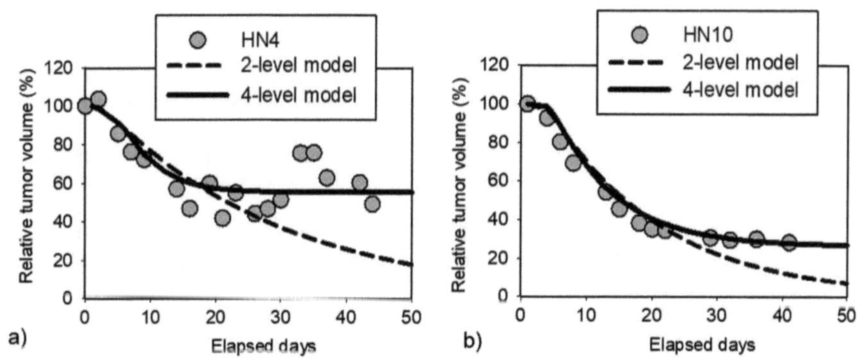

Fig. 2.10 Comparison of modeled time dependent tumor-volume variation during radiation therapy with measured data for two patients (HN10 and HN7)

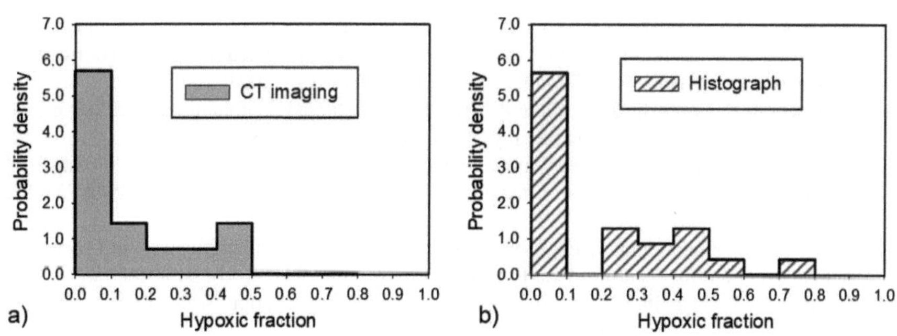

Fig. 2.11 Distribution of hypoxic fractions for head and neck cancer obtained from CT imaging data (**a**) and measured using histograph (**b**)

model of tumor response and serial volumetric imaging. This approach can be used for improved treatment response assessment and for guidance of more expansive PET measurements. Volumetric imaging is not expensive and is already integrated in many image-guided radiotherapy protocols; therefore, using measured tumor volume variation for evaluation of hypoxic fraction may present large opportunities for improved treatment response analysis.

References

1. Fischer JJ. Mathematical simulation of radiation therapy of solid tumors, I. Calculations. Acta Radiol Ther Phys Biol. 1971; 10(1):73-85.
2. Fischer JJ. Mathematical simulation of radiation therapy of solid tumors, II. Fractionation. Acta Radiol Ther Phys Biol. 1971; 10(2): 267-278.
3. Okumura Y, Ueda T, Mori T, Kitabatake T. Kinetic analysis of tumor regression during the course of radiotherapy. Strahlentherapie. 1977; 153(1): 35-39.
4. Lim K, Chan P, Dinniwell R, Fyles A, Haider M, Cho YB, Jaffray D, Manchul L, Levin W, Hill RP. Milosevic M. Cervical cancer regression measured using weekly magnetic resonance imaging during fractionated radiotherapy: Radiobiologic modeling and correlation with tumor hypoxia. Int J Radiat Oncol Biol Phys. 2008. 70(1): 126–133.
5. Huang Z, Mayr NA, Yuh WTC, Lo SS, Montebello JF, Grecula JC, Lu L, Li K, Zhang H, Gupta N, Wang JZ. Predicting outcomes in cervical cancer: A kinetic model of tumor regression during radiation therapy. Cancer Res. 2010; 70(2): 463-470.
6. Chvetsov AV, Yartsev S, Schwartz JL, Mayr N. Assessment of interpatient heterogeneity in tumor radiosensitivity for nonsmall cell lung cancer using tumor-volume variation data. Med. Phys. 2014; 41(6): 064101–1–10.
7. Chvetsov AV, Sandison GA, Schwartz JL, Rengan R. Ill-posed problem and regularization in reconstruction of radiobiological parameters from serial tumor imaging data. Phys Med Biol. 2015; 60(21): 8491-8503.
8. Unkelbach J, Craft D, Hong T, Papp D, Ramakrishnan J, Salari E, Wolfgang J, Bortfeld T. Exploiting tumor shrinkage through temporal optimization of radiotherapy. Phys Med Biol. 2014; 59(12): 3059-3079.
9. Ruggieri R, Nahum AE. The impact of hypofractionation on simultaneous dose-boosting to hypoxic tumor subvolumes. Med Phys 2006; 33(11): 4044-4045.
10. Jeong J, Shogi KI, Deasy JO. Modeling the interplay between hypoxia and proliferation in radiotherapy tumor response. Phys. Med. Biol. 2013; 58: 4897-4919
11. Dionysiou DD, Stamatakos GS, Uzunoglu NK, Nikita KS, Marioli A. A four-dimensional simulation model of tumor response to radiotherapy in vivo: parametric validation considering radiosensitivity, genetic profile and fractionation. J Theor Biol. 2004; 230:1-20.
12. Borkenstein K, Levegrün S, Peschke P. Modeling and computer simulations of tumor growth and tumor response to radiation therapy. Rad Res. 2004; 162:71-83.
13. Titz B, Jeraj R. An imaging-based tumor growth and treatment response model: investigating the effect of tumor oxygenation on radiation therapy response. Phys Med Biol. 2008; 53(17): 4471-4488.
14. Schwilden H, Honerkamp J, Elster C. Pharmacokinetic model identification and parameter estimation as an ill-posed problem. Eur J Clin Pharmacol. 1993; 45(6): 545-550.
15. Tikhonov AN, Arsenin VY. Solution of Ill-posed Problems (Scripta series in mathematics). Washington, DC: Winston & Sons; 1977.
16. Chvetsov AV, Dong L, Palta JR, Amdur RJ. Tumor-volume simulation during radiotherapy for head-and-neck cancer using a four-level cell population model. Int J Radiat Oncol Biol Phys. 2009; 75(2): 595-602.

17. Rodríguez-Romero R, Castro-Tejero P. The influence of respiratory motion on CT image volume definition. Med Phys. 2014; 41(4): 1-11.
18. Nawaratne S, Fabiny R, Brien JF, Zalcberg J, Cosolo W, Whan A,Morgan DJ Accuracy of volume measurement using helical CT J Comput Assist Tomogr. 1997; **21**(3): 481-486.
19. Rasey JS, Koh WJ, Evans ML, Peterson LM, Lewellen TK, Graham MM, Krohn KA. Quantifying regional hypoxia in human tumors with positron emission tomography of [^{18}F] fluoromisonidazole: A pretherapy study of 37 patients. Int J Radiat Oncol Biol Phys. 1996; 36(2): 417-428.
20. Wouters BG. Cell death after irradiation: How, when and why cells die? In: Basic Clinical Radiobiology. ed M Joiner and A. van der Kogel A. United Kingdom: Hodder Arnold; 2009.
21. Aoki T, Nagata Y, Negoro Y, Takayama K, Mizowaki T, Kokubo M, Oya N, Mitsumori M, Hiraoka M. Evaluation of lung injury after three-dimensional conformal stereotactic radiation therapy for solitary lung tumors: CT appearance. Radiology 2004; 230: 101-108.
22. Chvetsov AV, Palta JR, Nagata Y. Time-dependent cell disintegration kinetics in lung tumors after irradiation. Phys Med Biol. 2008; 53: 2413-2423.
23. Clarke G, Collins RA, Leavitt BR, Andrews DF, Hayden MR, Lumsten CJ, Mcinnes RR. A one-hit model of cell death in inherited neuronal degenerations. Nature 2000; 406: 195-199.
24. Koh WJ, Bergman KS, Rasey JS, Peterson LM, Evans ML, Graham MM, Grierson JR, Lindsley KL, Lewellen TK, Krohn KA, Griffin TW. Evaluation of oxygenation status during fractionated radiotherapy in human nonsmall cell lung cancer using (F-18) fluromisonidazole positron emission tomography. Int J Radiat Oncol Biol Phys. 1995; 33(2): 391-398.
25. Nakamura K, Brahme A. Evaluation of fractionation regimens in stereotactic radiotherapy using a mathematical model of repopulation and reoxygenation. Radiation Medicine 1999; 17(3): 219-225.
26. Carlson DJ, Stewart RD, Semenenko VA. Effects of oxygen on intrinsic radiation sensitivity: A test of the relationship between aerobic and hypoxic linear-quadratic LQ model parameters. Med Phys. 2006; 33(9):3105-3115.
27. Chvetsov AV, Rajendran JG, Zeng J, Patel SA, Bowen SR, Kim EY. Theoretical effectiveness of cell survival in fractionated radiotherapy with hypoxia-targeted dose escalation. Med Phys. 2017; 44(5):1975-1982.
28. Becker A, Hänsgen G, Bloching M, Weigel C, Lautenschläger C, Dunst J. Oxygenation of squamous cell carcinoma of the head and neck: comparison of primary tumors, neck node metastases, and normal tissue. Int J Radiat Oncol Biol Phys. 1998; 42(1):35-41.
29. Nordsmark M, Overgaard M, Overgaard J. Pretreatment oxygenation predicts radiation response in advanced squamous cell carcinoma of the head and neck. Radiother Oncol. 1996; 41: 31-39.
30. Tannock I, Howes A. The response of viable tumor cords to a single dose of radiation. Rad Res. 1973; 55: 477-486.
31. Byrd R H, Lu P, Nocedal J, Zhu C. A limited memory algorithm for bound constrained optimization. SIAM J Sci Comput. 1995; 16:1190-1208.
32. Shibamoto Y. Hara M. Radiobiology of normal lung tissue and lung tumors. In: Advances in radiation oncology in lung cancer, ed B Jeremić. Berlin: Springer Verlag; 2005.
33. Wigg DR. Applied Radiobiology and Bioeffect Planning. Madison, WI. Medical Physics Publishing; 2001.
34. Hansen PC. Analysis of discrete ill-posed problems by means of the L-curve SIAM Review 1992; 34(4): 561-580.
35. Wei J, Sandison GA, Chvetsov AV. Reconstruction of electron spectra from depth doses with adaptive regularization. Med Phys. 2006; 33(2): 354-359.
36. Chvetsov AV, Dempsey JF, Palta JR. Optimization of equivalent uniform dose using the L-curve criterion. Phys Med Biol 2007; 52: 5973-5984.
37. Barker JL, Garden AS, Ang KK, O'Daniel JC, Wang H, Court LE, Morrison WH, Rosenthal DI, Chao KS, Tucker SL, Mohan R, Dong L. Quantification of volumetric and geometric changes occurring during fractionated radiotherapy for head-and-neck cancer using an integrated CT/ linear accelerator system. Int J Radiat Oncol Biol Phys. 2004; 59(4): 960-970.

38. Corana A, Marchesi M, Martini C, Ridella S. Minimizing multimodal functions of continuous variables with the "simulated annealing" algorithm. ACM Trans Math Software. 1987; 13(3): 262-280.
39. Goffe WL, Ferrier GD, Rogers J. Global optimization of statistical functions with simulated annealing. J Econometrics 1994; 60(1/2): 65-99.

Chapter 3
Dose-Volume Effects in Tumor Control Probability

3.1 Dose-Volume Effects in TCP

3.1.1 Clinical Studies

Clinical studies in radiotherapy with conventional dose fractionation have shown a reduction in the TCP with increasing initial total tumor volume as measured using computed tomography (CT) or magnetic resonance imaging (MRI). The reduction in the TCP for the larger tumors has been reported for head and neck cancer, non-small cell lung cancer, melanoma, and cervical cancer [1–6]. More recent clinical studies based on functional imaging with PET or MRI have also shown a TCP reduction with increased hypoxic tumor volumes [7, 8]

Clinical studies in hypofractionated SBRT have also shown a reduction in the TCP with increasing initial tumor volume. For example, the reduction in the TCP in non-small cell lung cancer treated with SBRT with increasing initial volume has been reported for larger T2 lesions (tumor size = 3–7 cm) compared to smaller T1 lesions (tumor size < 3 cm) [9–11]. The local control probability is typically reported as > 90% for T1 lesions but decreases to 60–80% for T2 lesions. The influence of tumor size on treatment outcomes was summarized in [11] showing that when an equal radiation dose was delivered to T1 and T2 tumors, local control was inferior in T2 tumors compared with T1. A quantitative radiobiological mechanism for reduction of the TCP with increasing initial tumor volume in SBRT has not been formulated and thoroughly tested. However, there are several published studies in SBRT that demonstrate the dose escalation to the total tumor volume improves local control of larger T2 tumors [12–14].

A plausible mechanism for the volume effects in the TCP would be useful for analysis of treatment outcomes and design of dose escalation protocols. The dose-volume effects in the TCP represents one of the fundamental limitations of radiation therapy because it requires an application of larger doses to larger tumor volumes

A. V. Chvetsov, *Optimization Problems in Radiotherapy for Hypoxic Tumors*, Biological and Medical Physics, Biomedical Engineering,
https://doi.org/10.1007/978-981-96-7001-7_3

to maintain equal TCP. In radiotherapy with external beams, it leads to irradiation of larger volumes of normal tissue with higher radiation dose. Irradiation of larger volumes of normal tissue is associated with increased complications because of the dose-volume effects in normal tissues described in Sect. 1.6; thus, creating constraints for dose escalation. Therefore, it is important to characterize the volume effect in TCP quantitatively and understand its dependence on different radiobiological parameters.

3.1.2 TCP Equation

Equation 1.24 for the TCP predicts that a reduction in TCP with increasing initial tumor volume may be explained by several factors, including: (1) an increase in the initial number of clonogenic cells and (2) an increase in the average cellular radiation resistance of the tumor clonogenic cells. For well-oxygenated tumors, an increase in the initial number of clonogenic cells is the key model parameter that predicts a reduction in the TCP with increasing initial tumor volume. However, for hypoxic tumors, a TCP reduction with increasing tumor volume may be related to the combined effects of the initial number of clonogenic cells and the well-known increased cellular radio-resistance of cells irradiated under conditions of reduced oxygen described in Sect. 1.2.

One of the main objectives of this Chapter is to derive an analytical dependence of the TCP on the hypoxic and total tumor volumes that could be used to evaluate the hypoxia-targeted dose escalation. The derived equations may be used for evaluation of the patient specific radiation doses and analysis of treatment outcomes as well. The TCP equation predicts a reduction in the TCP with an increase in the initial number of clonogenic cells or the average cell surviving fraction. It is important to evaluate the contribution of these two processes with regards to TCP reduction because it may help in the development of more advanced radiotherapy approaches and analysis of clinical data. It would be difficult to separate these two effects in experimental or clinical studies; however, some estimates of the magnitude of these effects can be obtained through computational studies. To achieve this goal, we investigated the radiobiological factors that define the TCP dependence on the hypoxic and total tumor volume.

The TCP reduction due to increasing tumor volume and the total number of clonogenic contained therein has been examined for conventional fractionated radiotherapy treatments [1, 2]. In these earlier studies, TCP calculations were based on a tumor response model in which cellular response parameters are uniform, i.e., the α and β parameters in the LQ cell survival model are the same throughout the entire tumor volume. The numeric values for the LQ parameters, along with the initial number of clonogenic cells, are then typically determined from fits to clinical data. The estimates of cellular radiosensitivity given by the values of "α" in the LQ model in these studies were usually significantly lower than the values obtained from the in vitro predictive assays. This is an indication that the TCP reduction with the initial tumor volume may be a combined effect of the increase in the total number of clonogenic cells and mechanisms of cellular radioresistance. Therefore, we believe that it would be useful to evaluate potential mechanisms underlying the increased radiation resistance associated with the initial tumor volume (number of clonogenic cells) and further investigate the relationship of TCP and the number of clonogenic cells in representative tumor models that consider the effects of tumor heterogeneity.

3.2 TCP Models for Hypoxic Tumors

3.2.1 Two-Level TCP Model

To derive the dose-volume dependence of the TCP for hypoxic tumors, we use an approximation with two discrete levels of viable well-oxygenated and hypoxic clonogenic cells. This approximation stems from the 4-level tumor response model described in Chap. 2 with the omission of lethally damaged cells. The TCP is defined only by the number of viable clonogenic cells (n_1 and n_3) and the number of lethally damaged clonogenic cells (n_2 and n_4) can be omitted [15, 16]:

$$TCP = \exp(-\gamma n_{ox,0} - \gamma n_{hyp,0}), \tag{3.1}$$

where $n_{ox,0}$ and $n_{hyp,0}$ are the initial numbers of oxygenated and hypoxic cells in the tumor respectively and γ is the relative density of clonogenic cells. This approach has an advantage of minimizing the number of ad hoc adjustable parameters that need to be determined from clinical observations while still providing a formalism to explicitly investigate the relationship between two key mechanisms of action in the TCP modeling of tumor response. Several levels of pO_2 have been also used by other authors [17, 18].

To convert from the number of clonogenic cells to the tumor volumes we assume a power-law relationship between the initial tumor volumes and the number of clonogenic cells given by Eq. 1.26:

$$\gamma n_{l,0} = \rho V_{l,0}^{a}, \quad l = ox, hyp \tag{3.2}$$

where ρ is the density of clonogenic cells, $V_{j,0}$ is the initial hypoxic or well-oxygenated tumor volume and a is a volume exponent. We use the value of $a = 1$, i.e., the number of clonogenic is directly proportional to the initial tumor volume $V_{\mathrm{j},0}$.

Substituting Eq. 3.2 into Eq. 3.1 for the TCP and applying the equations of the 4-level tumor model for fractionated radiotherapy, we obtain that the overall TCP is the product of the TCP for oxygenated and hypoxic tumor sub-volumes:

$$TCP = \prod_{l} TCP_l, \quad l = ox, hyp, \tag{3.3}$$

where

$$TCP_{ox} = \exp\{-\rho V_{ox,0} S^k(d_{ox}) \exp((k-1)\lambda\Delta t)\} \tag{3.4}$$

and

$$TCP_{hyp} = \exp\{-\rho V_{hyp,0} S^k(d_{hyp}.OER)\} \tag{3.5}$$

The following notations are used in Eqs. 3.3–3.5: $S(d_{ox})$ and $S(d_{hyp}, OER)$ are the cell survival fractions for well-oxygenated and hypoxic cells for the average fractional doses d_{ox} and d_{hyp} in oxygenated and hypoxic volumes, respectively, k—is the number of dose fractions, λ is the cell proliferation constant and Δt is the time interval between dose fractions. For well-oxygenated oxygenated tumors, if the hypoxic volume $V_{\text{hyp},0} = 0$, Eq. 3.4 can be used for the TCP calculations. The final equation for the TCP in this case is given by:

$$\begin{aligned} &TCP = TCP_{ox} = \exp\{-\rho V_{tot,0} S^k(d_{ox}) \exp((k-1)\lambda\Delta t)\}, \\ &if \quad V_{tot,0} = V_{ox,0}. \end{aligned} \tag{3.6}$$

where $V_{\text{tot},0}$ is the total tumor volume.

In Eq. 3.3, we assume an idealized approximation that the hypoxic tumor cells are non-dividing or slowly dividing so that the rate of tumor repopulation in these parts of tumor is negligible. We also assume the worst case (most conservative) scenario for the TCP evaluation in which no re-oxygenation of hypoxic regions of the tumor occurs during the treatment. If the tumor re-oxygenation rate is known, the initial hypoxic volume in Eq. 3.3 can be corrected considering re-oxygenation rate A as discussed in Chap. 2 for the 4-level tumor model [19].

In practical applications, it is not feasible to compute the escalated dose using the re-oxygenation rate because it is a parameter that is not known prospectively and may change during the treatment. Therefore, we accepted an approximation that does not include re-oxygenation to have a worst-case scenario for TCP calculation. However, if the hypoxic fraction changes significantly during the treatment due to re-oxygenation, the escalated dose and volume could be recalculated using the same approach. The inclusion of re-oxygenation will not change the conclusions; however, the hypoxic volumes should be considered at the end of radiotherapy to consider the re-oxygenation rate We also note that impact of re-oxygenation is smaller for relatively shorter hypofractionated treatments.

3.2.2 *TCP for Large Hypoxic Volumes*

Equation 3.3 includes the TCP dependence on well-oxygenated and hypoxic tumor volumes. This dependence is complex; however, the specific form of cell survival function given by Eq. 1.15 allows one to simplify the equation for the TCP given by Eq. 3.3. Specifically, the cell surviving fraction in the oxygenated region is much smaller than that in the hypoxic region for the OER_{max} in the range $OER_{max} = 2.5–3.5$ as it was discussed in Chap. 1. Therefore, the following condition can be obtained for the ratio of cell surviving fractions:

$$\frac{S^k(d_{ox})}{S^k\left(d_{hyp}, OER\right)} = Q \ll 1. \tag{3.7}$$

The parameter Q in Eq. 3.7 is significantly smaller than 1.0 at the dose levels used in curative fractionated radiotherapy. However, parameter Q may approach 1.0 for the small values of the OER and the large hypoxia-targeted dose escalation in hypoxic region. The dependence of the parameter Q on the OER and the dose escalation in the hypoxic region is investigated below in this Chapter.

The TCP equation given by Eq. 3.3 can be simplified to Eq. 3.8 for the hypoxic tumors, assuming the hypoxic fraction is larger than a small threshold because we obtain the condition $TCP_{ox} \approx 1$ considering Eq. 3.7:

$$\begin{aligned} &TCP \approx TCP_{hyp} = \exp\left\{-\rho V_{hyp,0} S^k\left(d_{hyp}, OER\right)\right\}, \\ &if\, \frac{V_{hyp,0}}{V_{tot,0}} > Q. \end{aligned} \tag{3.8}$$

3.2.3 TCP for Intermediate Hypoxic Volumes

For intermediate fractional hypoxic volumes, the TCP can be calculated only complex Eq. 3.3. To evaluate the dependence of the TCP on intermediate hypoxic volumes, we compute the tumor average cell surviving fraction S_{TA} described in Chap. 1. To evaluate the average cell survival in a hypoxic tumor, we use an approximation that includes two levels of oxygenated and hypoxic cells that is described in Chap. 2. This approach is used primarily for mathematical brevity that allows us to obtain analytical solutions. Using the approach of the 2-level tumor model with oxygenated and hypoxic cells and assuming equal density of clonogenic cells in the hypoxic and oxygenated tumor regions, the tumor-average cell surviving fraction S_{TA} can be evaluated from the equation based on Eq. 1.29 obtained in Chap. 1:

$$S_{TA} = v_{ox} S^k(d_{ox}) + v_{hyp} S^k\left(d_{hyp}, OER\right), \tag{3.9}$$

where v_{ox} and v_{hyp} are the initial fractional hypoxic and total tumor volumes, $S(d_{ox})$ and $S(d_{hyp})$ are the cell surviving fractions for oxygenated and hypoxic cells,,and k is the number of dose fractions. Equation 3.9 presents a relationship between doses d_{ox}

and d_{hyp} and the fractional hypoxic volumes v_{ox} and v_{hyp} assuming that the average cell survival S_{TA} is given.

We computed the tumor-average cell surviving fraction S_{TA} given by Eq. 3.9 for non-small cell lung cancer using the parameters of the LQ model $\alpha = 0.45$ Gy^{-1} and $\beta = 0.039$ Gy^{-2} [20] for conventional dose regimen with total dose of 60 Gy in 30 fractions and hypofractionated SBRT dose regimen with total dose of 48 Gy in 4 fractions. Applying approximation of the 2-level tumor model, we use two discrete levels of OER: OER = 1 for well-oxygenated cells and $\text{OER}_{\text{max}} = 2$ for hypoxic cells for conventional fractionation and $\text{OER}_{\text{max}} = 3$ for hypofractionation. In Fig. 3.1, we show the variation of tumor-average cell survival S_{TA} as a function of relative hypoxic volume without dose escalation and with 50% dose escalation relative to the dose in the oxygenated region. One can see that the tumor-average cell surviving fraction in both model problems is defined by the cell survival in hypoxic region for a relative hypoxic volume larger than a threshold value of about 3%. We note that the tumor-average cell surviving fraction in hypofractionated dose regimens is smaller compared to conventional fractionation for oxygenated tumors; however, it becomes larger compared to conventional fractionation for hypoxic tumors because of larger value of the OER. These computational results support the conclusions who also obtained reduced tumor cell kill in hypofractionation compared to conventional fractionation for tumors with regions of hypoxia [21].

Using results of this simulation, it is easy to see that under conditions of moderate dose escalation assumed in this chapter the following condition is valid:

$$S^k(d_{ox}) \ll S^k\left(d_{hyp}, OER\right). \tag{3.10}$$

Therefore, Eq. 3.9 for tumor-average cell surviving fraction can be simplified to:

$$S_{TA} \approx v_{hyp} S^k\left(d_{hyp}, OER\right), \tag{3.11}$$

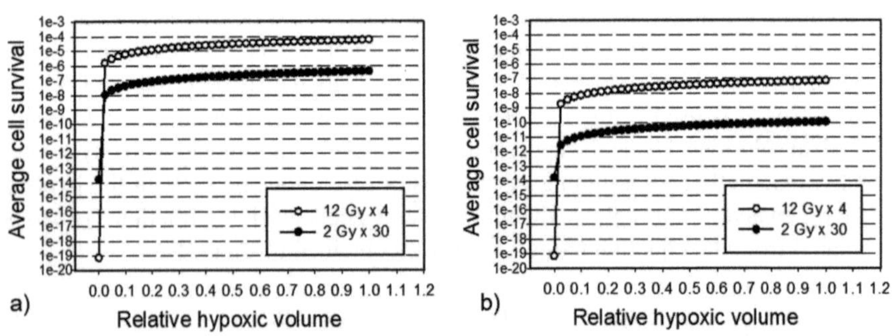

Fig. 3.1 Variation of average tumor cell survival as a function of relative hypoxic volume for conventional 2 Gy × 30 and hypofractionated dose regimens 12 Gy × 4: **a** model problem without dose escalation and **b** model problem with hypoxia-targeted dose escalation (50% of dose in oxygenated region)

if the fractional hypoxic volume is larger than a small threshold value of about 3%. We note that cell proliferation and re-oxygenation are not included in this analysis, but these approximations cannot be considered as a limitation of this approach. Proliferation was not included in our analysis because the TCP for hypoxic tumors depends primarily on hypoxic fraction that does not proliferate.

3.3 Dose-Volume Dependence of TCP

3.3.1 Dose-Volume Effects

The analytical dependence of dose on total and hypoxic volumes can be derived by substituting Eq. 1.15 for cell surviving fraction for oxygenated $S(d_{ox})$ and hypoxic $S(d_{hyp})$ cells in Eqs. 3.6 and 3.8 respectively. Applying standard solution to final quadratic equations, we obtain the analytical equation for dose as a function of total and hypoxic volumes:

$$d_l = OER_l \frac{-\alpha_{ox} + \sqrt{\alpha_{ox}^2 + 4\beta_{ox} C_l}}{2\beta_{ox}}, \quad l = ox, hyp, \tag{3.12}$$

where

$$C_{ox} = \frac{1}{k} \ln\left\{-\frac{\rho V_{tot,0}}{\ln TCP}\right\} + \frac{(k-1)\lambda \Delta t}{k}, \tag{3.13}$$

and

$$C_{hyp} = \frac{1}{k} \ln\left\{-\frac{\rho V_{hyp,0}}{\ln TCP}\right\}. \tag{3.14}$$

Equations 3.12 and 3.14 describing the dose-volume dependence of TCP for hypoxic tumors in conventional fractionation will be also valid for hypofractionation because we have shown in Chap. 1 that the LQ model is valid for hypoxic tumors at high dose doses. However, the dose-volume dependence of the TCP for well-oxygenated tumors needs to be corrected at high doses because the LQ model is not valid in this region. Using multitarget component of the USC [22] given by Eq. 1.16 to describe the cell surviving fraction $S\,(d_{ox})$ in the TCP equation for well-oxygenated tumors given by Eq. 3.6, we obtain the following dose dependence on the initial tumor volume for oxygenated tumors:

$$d_{ox} = D_0 C_{ox} + D_q, \tag{3.15}$$

where C_{ox} is defined by Eq. (3.13).

3.3.2 Derivatives of Dose-Volume Dependence

To compare the steepness of dose-volume dependence for oxygenated and hypoxic tumors, we computed and compared the derivative of these functions. The derivative of dose-volume dependence of the TCP for oxygenated tumors is given by:

$$\frac{\partial d_{ox}}{\partial V_{tot,0}} = \frac{1}{nV_{tot,0}\sqrt{\alpha_{ox}^2 + 4\beta_{ox}C_{ox}}}, \tag{3.16}$$

and the derivative of dose-volume dependence of the TCP for hypoxic tumors is given by

$$\frac{\partial d_{hyp}}{\partial V_{hyp,0}} = \frac{OER}{nV_{hyp,0}\sqrt{\alpha_{ox}^2 + 4\beta_{ox}C_{hyp}}}. \tag{3.17}$$

Computing a ratio of the derivatives of dose-volume dependence for oxygenated and hypoxic tumors given by Eqs. 3.16 and 3.17, respectively, we obtain that the derivative for hypoxic tumors is always larger by at least the OER than the derivative for oxygenated tumors:

$$\frac{\partial d_{hyp}}{\partial V_{hyp,0}}\left(\frac{\partial d_{ox}}{\partial V_{tot,0}}\right)^{-1} \geq OER. \tag{3.18}$$

Equation 3.18 reveals a larger volumetric dependence of the TCP if a tumor is hypoxic. This effect can be evaluated by computing the derivative:

$$\frac{d(TCP)}{dV_0} = -\rho S_{av} TCP \tag{3.19}$$

Equation 3.19 predicts that a larger cell surviving fraction and a larger density of clonogenic cells will produce steeper volume dependence of TCP.

3.4 TCP for Conventionally Fractionated Radiotherapy

3.4.1 Clinical Data and Radiotherapy Models

(a) **Radiotherapy for Head and Neck Cancer**

To evaluate the volume and dose-volume dependence of the TCP for conventional dose fractionation, we consider two radiotherapy model problems. The first problem was designed to mimic the conventional dose regimen in the clinical study for head and neck cancer [4]. The treatment schedule used in this study consists of a primary treatment delivered during six weeks and a concomitant boost delivered during the last three weeks at least six hours after primary dose fractions. The simulations were performed using a uniform dose fractionation schedule with equal time interval between dose fractions delivered during six weeks to accurately account for tumor repopulation over the course of treatment. Total cumulative median tumor dose in this study was 70.2 Gy. The fractionation schedule in the simulations included the total cumulative tumor dose 70.2 Gy delivered in 39 fractions.

A set of radiobiological parameters ($\alpha = 0.36$ Gy^{-1}, $\beta = 0.042$ Gy^{-2}, $\lambda = 0.12$ day^{-1}) representative of head and neck cancer (squamous cell carcinoma (SCC)) has been selected [20]. The density of clonogenic cells is another key parameter that is required for the TCP model. The clonogenic densities estimated from clinical data and measured from plating efficiencies of in vitro human tumors by several authors are summarized in [20]. For head and neck cancers, majority of estimates are in the range from 1.0E+6 to 1.0E+7 cells per cm^3 for tumor clonogenic cells. We selected a representative density of 1.0E+6 cells per cm^3 for clonogenic cells that corresponds to an effective (overall) cell density of 1.0E+8 cells per cm^3 and a clonogenic cellular fraction of 1%.

(b) **Radiotherapy for Non-small Cell Lung Cancer**

The second model problem was designed to simulate the clinical study in conventional radiation therapy for non-small cell lung cancer (NSCLC) [5]. This study reported a two-year local failure rate for Stage III NSCLC patients with gross tumor volumes (GTVs) of 100 cm^3 or higher and 200 cm^3 or higher treated with total doses in the range from 50 to 70 Gy. The fractionation schedule in our simulations included the total cumulative tumor dose of 60 Gy delivered in 30 fractions. The simulations were performed using a uniform dose fractionation schedule with an equal time interval of 1.4 days between dose fractions to consider weekends and to approximate the effects of tumor repopulation over the course of treatment. The following radiobiological parameters ($\alpha = 0.45$ Gy^{-1}, $\beta = 0.039$ Gy^{-2}, $\lambda = 0.1$ day^{-1}) have been selected for the NSCLC model [20]. A representative density of 1.0E+6 cells per cm^3 was selected for clonogenic cells that corresponds to an effective (overall) cell density of 1.0E+8 cells per cm^3 and a fraction of 1% for clonogenic cells.

3.4.2 Simulation and Comparison with Clinical Data

(a) **Criteria for Hypoxia TCP Dominance**

Equation 3.7 describes the ratio of cell surviving fractions in the hypoxic and well-oxygenated tumor regions. It is an important condition for deriving the TCP dependence on the hypoxic volume. We evaluated the limits of applicability of this equation as a function of the OER and dose escalation in radiotherapy problem for head and neck cancer. The ratio of cell surviving fractions in oxygenated and hypoxic regions as a function of the OER and the fractional dose escalation in the hypoxic region

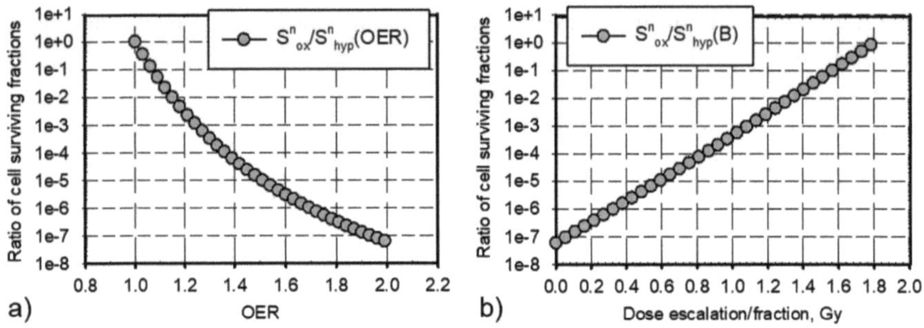

Fig. 3.2 Ratio of cell surviving fractions in oxygenated and hypoxic regions as a function of **a** OER in hypoxic region; **b** fractional dose escalation in hypoxic region for OER = 2 [15]

for $OER_{max} = 2$ is shown in Fig. 3.2a, b, respectively. One can see that the ratio of cell surviving fractions Q is smaller than 1.0E−4 in the range of OER = 1.4–2.0. The ratio Q is also smaller than 1.0E−4 for the dose escalation of less than 0.8 Gy per fraction. This is equal to 44% of the reference dose of 1.8 Gy in the oxygenated region. If we request Q < 1.0E−4, the described range of the OER and dose escalation defines the applicability of the threshold TCP equation.

(b) Comparison with Clinical Data

Figure 3.3 shows a comparison of clinical data for volume dependence of probability of recurrence (1-TCP) for head and neck cancer with the model predictions using oxygenated and hypoxic tumors. The calculations were performed with the hypoxic fractional volume equal to 20% (parameter $V_{hyp,0}/V_{tot,0} = 0.2$ in Eq. 3.8 and OER = 1.75. A comparison of the model predicted volume dependence of the probability of recurrence (1-TCP) for NSCLC cancer with the clinical data is shown in Fig. 3.4. The calculation for NSCLC was performed with a 40% hypoxic fractional volume and OER = 1.7. The FMISO PET measurements have been used as a guideline in selection of the hypoxic fractional volumes in NSCLC and head and neck cancers in our simulations [23].

As illustrated in Figs. 3.3 and 3.4, the dependence of probability of recurrence (1-TCP) on the initial tumor volume computed using oxygenated tumor model cannot describe the clinical data. However, a consideration of hypoxic volume brings the predicted trends with increasing initial tumor volume into closer agreement with the clinical data. The area under the (1-TCP) curves was used as a criterion of agreement between computed and clinical curves.

(c) Simulation of Dose-Volume Dependence of TCP

The dose-volume dependence of the TCP in radiotherapy model problem for head and neck cancer is shown in Fig. 3.5. The computations are performed using Eqs. 3.9–3.11 with a sample value of TCP = 0.9 and different values of the OER corresponding to well-oxygenated (OER = 1) and hypoxic (OER > 1) tumors. As shown in Fig. 3.4, the computed total dose for well-oxygenated tumors is smaller than the clinically

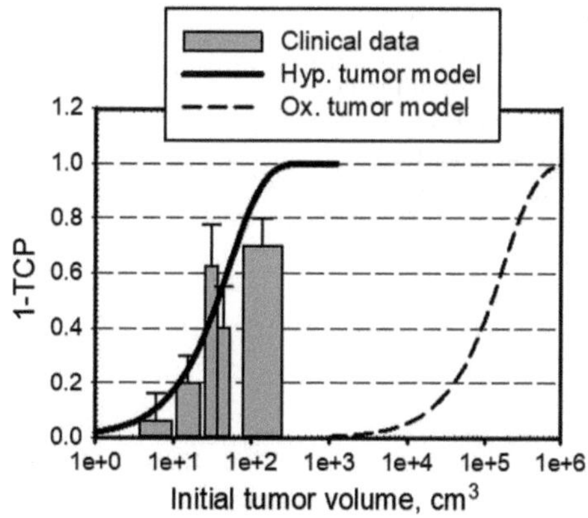

Fig. 3.3 Comparison of computed volume dependence of the probability of recurrence 1-TCP in the head and neck cancer radiotherapy model with clinical data of Johnson et al. [3]. The computed data are obtained using TCP models for oxygenated and hypoxic tumors [15]

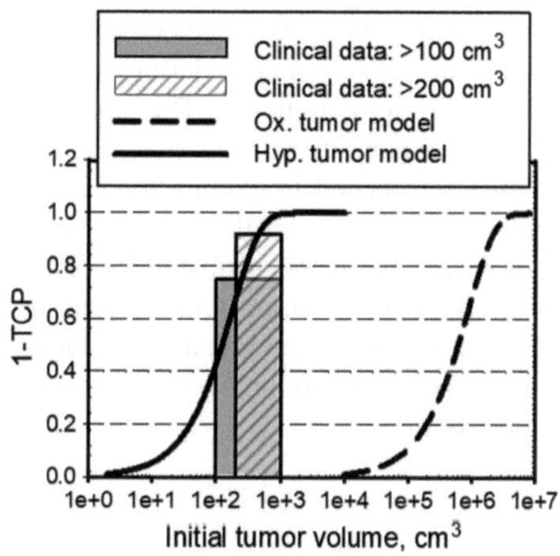

Fig. 3.4 Comparison of computed volume dependence of the probability of recurrence 1-TCP in the non-small cell lung cancer model problem with clinical data of Rengan et al. [5]. The computed data are obtained using TCP models for oxygenated and hypoxic tumors [15]

used prescription dose of 70.2 Gy for the entire computed range of volumes from 1 to 500 °C. This implies that predicted volume effect in the TCP is small for well-oxygenated tumors. The computed dose increases with increasing values of the OER for any given tumor volume and may exceed the prescribed dose of 70.2 Gy for hypoxic tumors (OER > 1). For example, the computed dose exceeds 70.2 Gy at 0.2 cm^3 for an OER = 2 and at 24 cm^3 for an OER = 1.5; therefore, the volume effect in the TCP may be larger for hypoxic tumors.

3.4.3 Conclusions

Our calculations show that, at the dose levels used in conventional radiotherapy, the TCP dependence on total tumor volume for completely oxygenated tumors is negligible compared to the TCP reduction observed in clinical studies. Therefore,

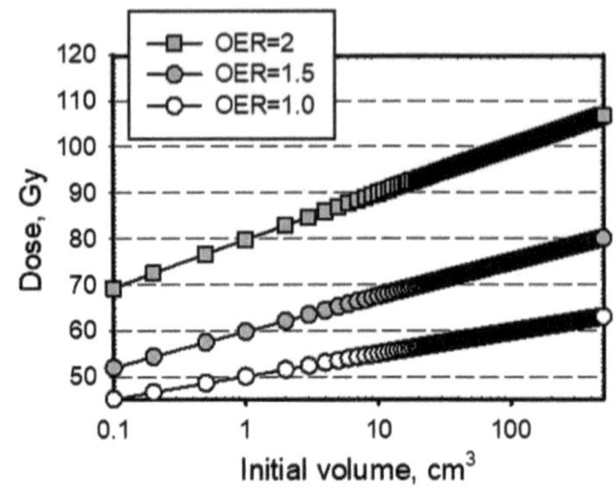

Fig. 3.5 Dose-volume dependence of TCP = 0.9 computed for different values of OER in the model problem describing conventional radiotherapy in head and neck cancer. Total dose of 70.2 Gy used in radiotherapy for head and neck cancer is shown [15]

the TCP reduction for larger tumor volumes observed in radiotherapy with conventional fractionation cannot be described by a change in the number of initial tumors clonogenic cells even considering possible perturbations about the mean in the LQ radiosensitivity parameters. This observation implies that the mechanisms of radiation resistance other than a change in the numbers of tumor cells with increasing volume are needed to explain the effect s of TCP reduction observed in clinical studies. An increase in the level of tumor hypoxia with increasing tumor volume is one plausible mechanism for the observed increase in radiation resistance with increasing tumor volume. Our simulations show that the volume dependence in the TCP is determined predominantly by the hypoxic volume. Therefore, the TCP reduction can be described largely by increasing the hypoxic tumor volume with the total tumor volume.

3.5 TCP for Hypofractionated Radiotherapy

3.5.1 Clinical Study and Radiotherapy Model

We have shown that an increase of hypoxic fraction with increasing the initial tumor volume provides a plausible explanation for the TCP reduction in radiotherapy with conventionally fractionated radiation therapy. In this section, we seek to test the same hypothesis for hypofractionated radiotherapy and demonstrate that the decrease in the TCP with increasing the initial tumor volume predominantly arises from an increase in the hypoxic region rather than an increase in the total number of tumors clonogenic cells. Comparative studies of the tumor response to conventional and hypofractionated radiation therapy are a potentially useful way to uncover and test hypothesized mechanisms of tumor resistance and to design more effective dose escalation protocols. For example, the standard LQ cell survival model may be inaccurate for larger dose per fraction treatments and the accurate modeling of local control in hypofractionated radiotherapy may require the use of alternate cell survival models such as USC described in Sect. 1.4. The effective OER may also be larger than the OER for conventional radiation therapy treatment schedules [16, 21]. Additionally, the hypofractionated treatments usually involve tumors with smaller volumes that may have smaller hypoxic fractions, but the short overall treatment time limits the opportunity for re-oxygenation which may counteract the smaller hypoxic volumes in SBRT.

A clinical study in hypofractionated radiotherapy for NSCLC for selected for simulation of volume effects in the TCP and comparison with clinical data [10]. One of the main goals of this study was evaluation the difference in the TCP for T1 tumors (tumor size < 3 cm) and T2 tumors (tumor size = 3–7 cm). The median PTV was 39.8 ml (range 7.98–47.0 ml) for T1 tumors and 102 ml (range 40.5–196.0 ml) for T2 tumors. Using published radiosensitivity parameters and density of clonogenic cells derived from predictive assays, we analyze the TCP data obtained in this clinical study for SBRT for NSCLC and determine the dominant factors of the TCP reduction with increasing the initial tumor volume.

A test model was designed to simulate the selected clinical study in hypofractionated radiotherapy for NSCLC. The fractionation schedule in the SBRT model is based on total cumulative tumor dose of 60 Gy delivered in 3 fractions with equal time interval of 2.0 days between dose fractions. The following radiobiological parameters of the USC and LQ cell survival curve for the NSCLC have used: $\alpha = 0.33\ \text{Gy}^{-1}$, $\beta = 0.038\ \text{Gy}^{-2}$, $D_0 = 1.25$ Gy, $D_q = 1.80$ Gy, and $D_T = 6.2$ Gy [22]. The cell proliferation constant $\lambda = 0.1\ \text{day}^{-1}$ was taken from [20]. The TCP calculations using Eqs. 3.6 and 3.8 require knowledge of ρ, the density of clonogenic cells. Several authors estimated clonogenic densities from clinical data and plating efficiencies of in vitro human tumors and this data was summarized in [20]. Using this data, a representative density of clonogenic cells of 1.0E+6 cells per cm^3 was selected for our TCP calculations in the SBRT model problem. This density of clonogenic cells corresponds to an effective overall cell density of 1.0E+8 cells per cm^3 and a fraction of clonogenic cells of 1%.

3.5.2 Comparison of TCP Model with Clinical Data

(a) **Criteria for TCP equation**

The ratio of cell surviving fractions in oxygenated and hypoxic regions as a function of the OER and the fractional dose escalation factor in hypoxic region with $\text{OER}_{max} = 3$ is shown in Fig. 3.6a, b, respectively. We evaluated the applicability of this ratio as a function of OER and the dose escalation in the NSCLC model using the USC equation. One can see that the ratio of cell surviving fractions is smaller than 1.0E−6 in the range 1.4–3.0 of the OER. This ratio is also smaller than 1.0E−7 for the dose escalation of less than 20 Gy per fraction. This dose escalation is equal to 100% of reference dose of 20 Gy in the oxygenated region. These results confirm that Eq. 3.8

is valid in the wide range of the dose and OER variation that provides a condition of applicability of the TCP dependence only on hypoxic volume.

(b) **Comparison with Clinical Data**

A comparison of computed probability of recurrence (1-TCP) on tumor volume with clinical data for SBRT for NSCLC is shown in Fig. 3.7. The probability of recurrence (1-TCP) was computed using the LQ and the USC models for well-oxygenated tumors and hypoxic tumors. As illustrated in Fig. 3.7a, the dependence of probability of recurrence on the initial tumor volume computed using oxygenated tumor model cannot describe the clinical data. The USC brings the calculations closer to the measured data; however, the discrepancy is still significant.

In Fig. 3.7b, we show the simulation for a hypoxic tumor. These calculations are performed with fractional volume hypoxic equal to 20% (parameter $V_{hyp,0}/V_{tot,0}$ =

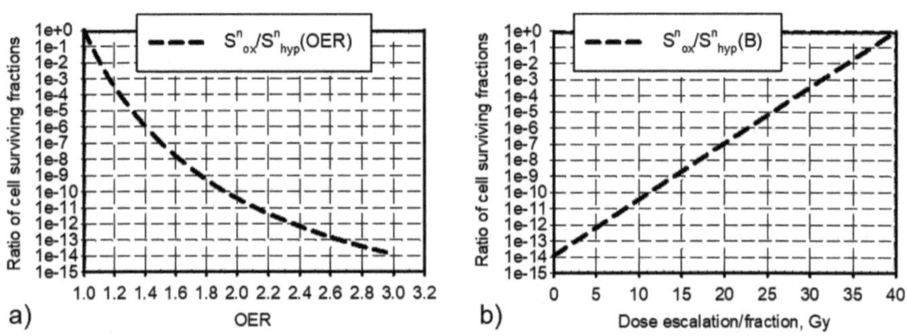

Fig. 3.6 The ratio of cell surviving fractions in oxygenated and hypoxic regions as a function of **a** OER in hypoxic region; **b** fractional dose escalation in hypoxic region with OER = 3 [16]

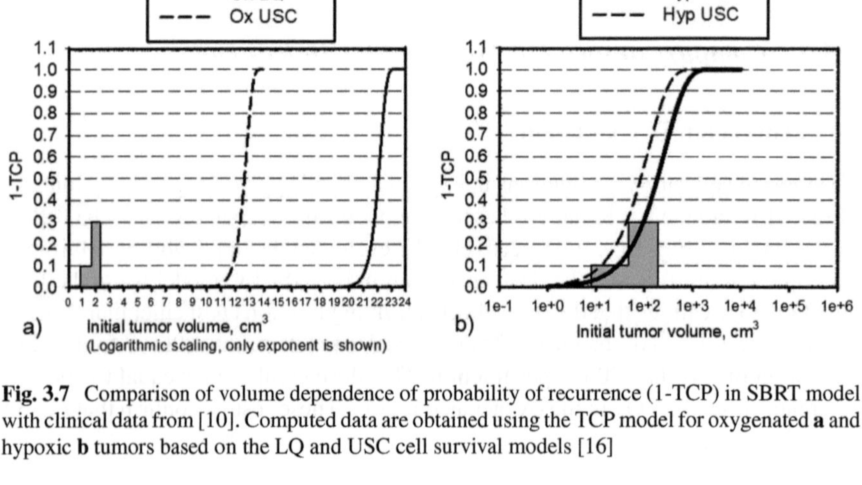

Fig. 3.7 Comparison of volume dependence of probability of recurrence (1-TCP) in SBRT model with clinical data from [10]. Computed data are obtained using the TCP model for oxygenated **a** and hypoxic **b** tumors based on the LQ and USC cell survival models [16]

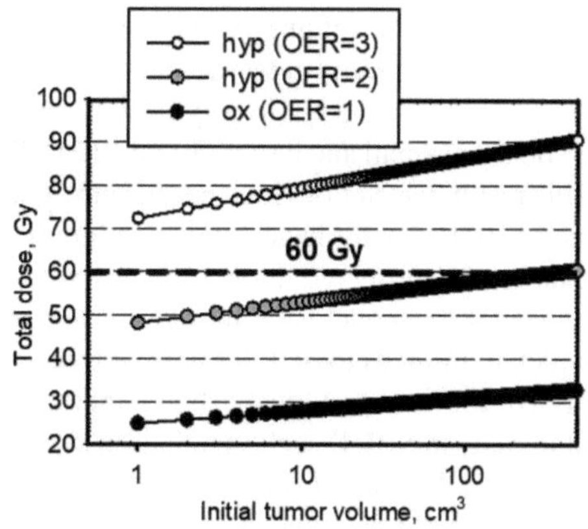

Fig. 3.8 Dose-volume dependence of the TCP = 0.8 computed for different values of OER in the SBRT model. Total dose prescription of 3 × 20 cGy = 60 Gy used in SBRT for NSCLC is shown [16]

0.2 in Eq. 3.8 and $OER_{max} = 2.25$. The 20% hypoxic fractional volume in relatively smaller tumors treated with SBRT was selected using the dependence of the hypoxic volume on total tumor volume in NSCLC obtained using FMISO PET measurements [23]. The consideration of a hypoxic volume brings the predicted trends in the TCP with increasing initial tumor volume into closer agreement with the clinical data. The agreement between computed and clinical curves was evaluated using the area under the (1-TCP) curves.

(c) **Dose-Volume Dependence of TCP**

Equations (3.6) and (3.8) can be used for evaluation of the TCP dependence on the initial total tumor and hypoxic volumes for a given dose fractionation regimen. It is also useful to validate the equations that describe the dependence of the radiation dose on the initial tumor and hypoxic volumes under condition of a constant value of the TCP. The dose-volume dependence of the TCP in the NSCLC radiotherapy model is shown in Fig. 3.8. The computations are performed with a sample value of TCP = 0.9 using Eqs. 3.13 and 3.15 for well-oxygenated tumors and Eqs. 3.12 and 3.14 for hypoxic tumors with different values of the OER. As shown in Fig. 3.8, the computed total dose for oxygenated tumors is smaller than the clinically used prescription dose of 60 Gy for the entire computed range of volumes from 1 to 400 cc. Similar results are obtained for hypoxic tumors with OER < 2. This implies that predicted volume effect in the TCP will be small for oxygenated tumors and hypoxic tumor with OER < 2. The volume effect can be observed in SBRT if the OER exceeds the value of OER = 2.

3.5.3 *Conclusions*

There are two main conclusions which can be drawn from the simulations presented in this Chapter. First, we have shown that the reduction in the TCP with increasing the initial tumor volume in hypofractionated SBRT can be explained by the extent of hypoxia. A trend towards increasing levels of hypoxia with increasing tumor volume can be expected to play a significant role in determining the different radiation response of small and larger (T1 and T2) NSCLC lesions observed in clinical studies. This result was not obvious because the tumors currently treated with hypofractionated SBRT usually have smaller volumes and may have smaller hypoxic volumes than usually observed in radiotherapy with conventional fractionation. The practical impact of this conclusion may result in optimized hypoxia targeted dose escalation which minimizes toxicity of normal tissues compared to the total volume escalation in SBRT. It was previously shown that tumor size correlates with the increased risk of toxicity in SBRT, such as radiation pneumonitis, fibrosis, and chest wall pain [24]; therefore, the correct selection of target volume is important.

Our second conclusion is that the LQ model can be used for TCP modeling in hypofractionated radiotherapy because the TCP is defined primarily by hypoxic cells. The larger value of OER in hypofractionated radiotherapy extends the applicability of the LQ model to larger doses. The usefulness of the LQ model for the TCP modeling in SBRT has also been shown in [25–27].

We have used several approximations in our calculations. First, we assumed slow re-oxygenation of hypoxic region during treatment that corresponds to the worst case (most conservative) scenario for the TCP evaluation. If the tumor re-oxygenation rate is known, the initial hypoxic volume in Eq. 3.8 can be corrected by taking re-oxygenation into account. Second, the dose distributions in GTV (Gross Tumor Volume) in SBRT are non-homogeneous with higher doses occurring at the center of the tumor. For oxygenated tumors, this effect would increase the equivalent doses to GTV and, consequently, the discrepancy with the clinical data. For hypoxic tumors, the location of the hypoxic region is not known, and this effect of dose non-uniformity was implicitly included in the effective value of OER.

References

1. Bentzen SM, Overgaard J, Thames HD, Overgaard M, Hansen PV, Maase H, Meder J. Clinical radiobiology of malignant melanoma. Radioth Oncol. 1989; 16(3):169–182.
2. Brenner DJ. Dose, volume, and tumor-control predictions in radiotherapy Int J Radiat Oncol Biol Phys. 1993; 26 171–179.
3. Bentzen SM, Thames HD. Tumor volume and local control probability: Clinical data and radiobiological interpretations. Int J Radiat Oncol Biol Phys. 1996; 36:247–251.
4. Johnson CR, Thames HD, Huang DT, Schmidt-Ulrich RK. The tumor volume and clonogen number relationship: Tumor control predictions based upon tumor volume estimates derived from computed tomography. Int J Radiat Oncol Biol Phys. 1995; 33(2):281–288.
5. Rengan R, Rosenzweig KE, Venkatraman E, Koutcher LA, Fox JL, Nayak R, Amols H, Yorke E, Jackson A, Ling CC, Leibel SA. Improved local control with higher doses of radiation in large-volume stage III non-small-cell lung cancer. Int J Radiat Oncol Biol Phys. 2004; 60:741–747.
6. Huang Z, Mayr NA, Gao M, Lo SS, Wang JZ, Jia G, Yuh WT. Onset time of tumor repopulation for cervical cancer: first evidence from clinical data. Int J Radiat Oncol Biol Phys. 2012;84(2):478–484.

7. Spence AM, Muzi M, Swanson KR, O'Sullivan F, Rockhill JK, Rajendran JG, Adamsen TC, Link JM, Swanson PE, Yagle KJ, Rostomily RC, Silbergeld DL, Krohn KA. Regional hypoxia in glioblastoma multiforme quantified with [18F] fluoromisonidazole positron emission tomography before radiotherapy: correlation with time to progression and survival. Clin Cancer Res. 2008; 14:2623–30.
8. Mayr NA, Huang Z, Wang JZ, Lo SS, Fan JM, Grecula JC, Sammet S, Sammet CL, Jia G, Zhang J, Knopp MV, Yuh WT. Characterizing tumor heterogeneity with functional imaging and quantifying high-risk tumor volume for earlier prediction of treatment outcome: cervical cancer as a model. Int J Radiat Oncol Biol Phys. 2012; 83(3):972–979.
9. van der Voort van Zyp NC, Prévost JB, Hoogeman MS, Praag J, van der Holt B, Levendag PC, van Klaveren RJ, Pattynama P, Nuyttens JJ. 2009 Stereotactic radiotherapy with real-time tumor tracking for non-small cell lung cancer: clinical outcome. Radiother. Oncol. 2009; 91: 296–300.
10. Dunlap NE, Larner JM, Read PW, Kozower BD, Lau CL, Sheng K, Jones DR. Size matters: A comparison of T1 and T2 peripheral non–small-cell lung cancers treated with stereotactic body radiation therapy (SBRT). Journal of Thoracic and Cardiovascular Surgery 2010; 140: 583–589.
11. Chi A, Liao Z, Nguyen NP, Xu J, Stea B, Komaki R. Systemic review of the patterns of failure following stereotactic body radiation therapy in earlier-stage non-small-cell lung cancer: Clinical implications. Radiother Oncol. 2010; 94: 1–11.
12. Zhao L, West BT, Hayman JA, Lyons S, Cease K, Kong FM. High radiation dose may reduce the negative effect of large gross tumor volume in patients with medically inoperable early-stage non-small cell lung cancer. Int J Radiat Oncol Biol Phys. 2007; 68: 103–110.
13. Trakul N, Chang CN, Harris J, Chapman C, Rao A, Shen J, Quinlan-Davidson S, Filion EJ, Wakelee HA, Colevas AD, Whyte RI, Dieterich S, Maxim PG, Hristov D, Tran P, Le QT, Loo BW Jr, Diehn M. Tumor volume-adapted dosing in stereotactic ablative radiotherapy of lung tumors. Int J Radiat Oncol Biol. Phys. 2012; 84: 231–237.
14. Allibhai Z, Taremi M, Bezjak A, Brade A, Hope AJ, Sun A, Cho BC. The impact of tumor size on outcomes after stereotactic body radiation therapy for medically inoperable early-stage non-small cell lung cancer. Int J Radiat Oncol Biol Phys. 2013; 87: 1064–70.
15. Chvetsov AV, Stewart RD, Kim M, Meyer J, Rengan R. Volume effects in the TCP for hypoxic and oxygenated tumors. Med Phys. 2020; 47: 4626–4633.
16. Chvetsov AV, Hanin LG, Stewart RD, Zeng J, Rengan R, Lo SS. Tumor control probability in hypofractionated radiotherapy as a function of total and hypoxic tumor volumes. Phys Med Biol. 2021; 66: 125010.
17. Daşu A, Toma-Daşu I, Karlsson M. The effects of hypoxia on the theoretical modelling of tumour control probability. Acta Oncol. 2005; 44(6): 563–71.
18. Malinen E, Sovik A, Hristov D, Bruland ØS, Olsen DR. Adapting radiotherapy to hypoxic tumors. Phys Med Biol. 2006; 51: 4903–4921.
19. Koh WJ, Bergman KS, Rasey JS, Peterson LM, Evans ML, Graham MM, Grierson JR, Lindsley KL, Lewellen TK, Krohn KA, Griffin TW. Evaluation of oxygenation status during fractionated radiotherapy in human nonsmall cell lung cancer using (F-18) fluromisonidazole positron emission tomography. Int J Radiat Oncol Biol Phys. 1995; 33(2): 391–398.
20. Wigg DR. Applied Radiobiology and Bioeffect Planning. Madison, WI. Medical Physics Publishing; 2001.
21. Carlson DJ, Keall PJ, Loo BW, Chen ZJ, Brown JM. Hypofractionation results in reduced tumor cell kill compared to conventional fractionation for tumors with regions of hypoxia. Int J Radiat Oncol Biol Phys, 2011; 74: 1188–1195.
22. Park C, Papiez L, Zhang S, Story M, Timmerman RD. Universal survival curve and single fraction equivalent dose: useful tools in understanding potency of ablative radiotherapy. Int J Radiat Oncol Biol Phys. 2008; 70: 847–852.
23. Rasey JS, Koh WJ, Evans ML, Peterson LM, Lewellen TK, Graham MM, Krohn KA. Quantifying regional hypoxia in human tumors with positron emission tomography of [^{18}F] fluoromisonidazole: A pretherapy study of 37 patients. Int J Radiat Oncol Biol Phys. 1996; 36(2): 417–428.

24. Zhao J, Yorke ED, Li L, Kavanagh BD, Li XA, Das S, Miften M, Rimner A, Campbell J, Xue J, Jackson A, Grimm J, Milano MT. Spring Kong FM. Simple Factors Associated with Radiation-Induced Lung Toxicity After Stereotactic Body Radiation Therapy of the Thorax: A Pooled Analysis of 88 Studies. Int J Radiat Oncol Biol Phys. 2016; 95: 1357–1366.
25. Ruggieri R, Stavreva N, Naccarato S. Stavrev P. Applying a hypoxia-incorporating TCP model to experimental data on rat sarcoma. Int J Radiat Oncol Biol Phys. 2012; 83: 1603–1608.
26. Liu F, Tai A, Lee P, Biswas T, Ding GX, El Naqa I, Grimm J, Jackson A, Kong FS, LaCouture T, Loo B Jr, Miften M, Solberg T, Li XA. Tumor control probability modeling for stereotactic body radiation therapy of early-stage lung cancer using multiple bio-physical models. Radiother Oncol. 2017; 122: 286–294.
27. Paz AE, Yamamoto N, Sakama M, Matsufuji N, Kanai T. Tumor Control Probability Analysis for Single-Fraction Carbon-Ion Radiation Therapy of Early-Stage Non-small Cell Lung Cancer. Int J Radiat Oncol Biol Phys. 2018; 102: 1551–1559.

Chapter 4
Equivalent Uniform Aerobic Dose

4.1 Bioeffect of Inhomogeneous Dose Distributions

4.1.1 Equivalent Uniform Dose (EUD)

Analysis and optimization of radiotherapy based on the TCP poses a challenging task due to large uncertainties in radiobiological parameters, for example, the density of tumor clonogenic cells that is usually unknown. Therefore, the concept of equivalent uniform dose (EUD) that is an intermediate concept between the physical dose and the TCP can be considered as a more robust and practical approach for optimization of biological effects. Dose distributions in radiation therapy are usually inhomogeneous in the target volumes and normal tissues. Therefore, assuming the same prescribed dose, a variation in the pattern of dose non-uniformity between different patients produces a variation in radiobiological response. It complicates the treatment outcome analysis and comparison of different treatment plans.

The EUD in radiation therapy is defined as the uniform dose that causes equal survival of clonogenic cells for a given non-uniform dose distribution [1, 2]. The concept of EUD assumes the conservation of the survival fraction of clonogenic cells for different dose distributions. The EUD was developed to introduce a single parameter for comparison and analysis of radiobiological effects for complex 3D distribution. The concept of EUD is a more accurate approach for analysis of heterogeneous dose distributions compared to other approaches such as the Effectively Delivered Dose [3, 4] because it considers the non-linearity of radiotherapy response. The Effectively Delivered Dose is calculated using the mean dose and the standard deviation which are the first and second moments of the non-uniform dose distribution, and cannot accurately take the non-linearity of radiation response into account. Apparently, the first application of the EUD called survival average dose was proposed for non-uniform dose distributions in therapeutic nuclear medicine [5]. This model utilized the LQ model to compute a uniform dose distribution producing

A. V. Chvetsov, *Optimization Problems in Radiotherapy for Hypoxic Tumors*, Biological and Medical Physics, Biomedical Engineering,
https://doi.org/10.1007/978-981-96-7001-7_4

equal cell surviving fraction for a non-uniform dose distribution. The concept of EUD for tumors in external beam radiotherapy was proposed in [1].

The concept of EUD has proven instrumental for the analysis of treatment outcomes and biological optimization of radiotherapy. Dose distributions in radiation therapy are usually inhomogeneous in the target volumes and normal tissues. Each treatment plan has a certain pattern of dose non-uniformity for equivalent dose prescription that produces a certain individual radiobiological response. Therefore, assuming the same prescribed dose, a variation in the pattern of dose non-uniformity between different patients produces a variation in radiobiological response. This complicates the treatment outcome analysis and comparison of different treatment plans.

A mathematical version of the EUD called a generalized EUD (gEUD) has found a broad application in radiotherapy treatment planning because of its parsimonious one-parametric nature [6]. The gEUD is widely used for evaluation and optimization of treatment plans, specifically, to optimize the EUD in inverse treatment planning for intensity modulated radiotherapy (IMRT) [7–9]. We also note the gEUD equation was introduced for computing the NTCP before it was introduced for tumors as it was described in Chap. 1 and was also used in the dose optimization algorithms [10].

The biologic or clinical validity of the EUD is defined largely by the concept that tumor control is a function of the minimal tumor dose with small dependence on the "hot spots" to the tumor. The concept of EUD assumes the conservation of the survival fraction of clonogenic cells. However, several authors noted that the EUD optimization is limited to the dose domain and is unable to predict the biologic response as opposed to the TCP-based models [11].

4.1.2 EUD in Hypoxic Tumors

The radiotherapy treatment planning concepts such as the TCP and EUD could serve as effective prognostic parameters for hypoxic tumor because they include dose regimens and non-linearity of radiotherapy response as well. In the analysis of radiotherapy outcomes, hypoxia has been usually characterized using the imaging derived parameters such as hypoxic volume, maximum uptake of hypoxia biomarker or more advanced radiomic features [12–16]. These imaging derived hypoxia parameters do not include the radiotherapy dose regimens, and the non-linearity of radiation response described by the LQ cell survival model. Therefore, the imaging-based hypoxia parameters may have limited prognostic value and cannot be used effectively in dose optimization where a relationship between treatment outcomes and three-dimensional dose distributions is needed. This limitation becomes more important when different hypofractionation dose regimens and SBRT are investigated for different cancers. The TCP and EUD include both radiation dose regimens and hypoxia features and could serve as a prognostic parameter for heterogeneous hypoxic tumors and the criterion for biological optimization.

However, the application of the TCP involves application of significant number of radiobiological parameters and the concept of EUD was developed for homogeneously oxygenated tumors; therefore, it is not applicable to heterogeneous hypoxic tumors which represent the majority of those encountered in humans. Most applications of EUD assume homogeneously oxygenated tumors that were used in most applications except [17]. The concept of EUD has a potential to be effective prognostic factor in radiotherapy for heterogeneous hypoxic tumors; however, its implementation will require significant research and testing. A more general concept of EUD is discussed in this book that is applicable to both hypoxic and oxygenated tumors.

First, the EUD does not consider the complex 3D distribution of radiosensitivity in tumors; therefore, it cannot be used for the comparison of radiobiological effects in tumors with intra- and inter-tumor variations in radiosensitivity. Second, the survival of clonogenic cells in hypoxic tumors is a function of both non-uniformity of dose distribution and non-uniformity the OER that is a function of partial oxygen pressure given by Eq. 1.2. Computing the EUD in radiotherapy for heterogeneous hypoxic tumors requires knowledge of the OER within the tumor that can be evaluated using PET CT or MR functional imaging. The PET hypoxia images present a distribution of SUV; therefore, the SUV needs to be converted to partial oxygen pressure pO_2 and then the partial oxygen pressure pO_2 can be converted to the OER as it was discussed in Chap. 1. The verification and testing the algorithms to derive the OER from functional imaging is needed.

Therefore, a more general definition of the EUD should be developed that incorporates the oxygen effect and could serve as a predictor of treatment outcomes in hypoxic tumors. There are several significant and practical applications for the EUD for heterogeneous hypoxic tumors in radiation therapy. First, the EUD can be used in treatment outcomes analysis and comparison effectiveness of different treatment plans for different dose regimens including hypofractionated regimens. Second, the EUD for heterogeneous hypoxic tumors can be used in patient-specific biological optimization because maximization of the EUD allows one to find optimal non-uniform dose distributions [17]. Progress in this research would not be possible without the definition of EUD in heterogeneous hypoxic tumors with highly heterogeneous dose distributions.

4.2 Derivation of Equivalent Uniform Aerobic Dose

4.2.1 Integrated Cell Survival

If the cells with a specific value of OER are irradiated with the uniform dose $d(pO_2)$ the equivalent dose under aerobic conditions d_{aer} that produces equivalent cell survival can be obtained from Eq. 1.3:

$$d_{aer} = \frac{d(pO_2)}{OER}. \tag{4.1}$$

The aerobic dose d_{aer} can be used as a parameter for comparison of cell survival for different combinations of physical dose d_{hyp} and the OER. The distributions of dose and OER in radiation therapy for malignant tumors are not uniform; however, it would be also useful to define an equivalent uniform dose for hypoxic tumors that would produce equal survival of clonogenic cells. This equivalent uniform dose could be used for radiotherapy optimization, treatment outcomes analysis, and comparison of the effectiveness of different treatment plans.

The computation of EUD in hypoxic tumors requires a special approach because both the radiation dose and the OER vary within a tumor. To derive equations for the EUD, we consider a tumor model separated into I small voxels and irradiated according to a fractionation schedule comprising of k fractions. Each voxel is characterized by the voxel specific OER and is irradiated with a fractional dose d_{i}. The OER and the dose can vary in the tumor, and we compute a uniform dose that produces equal radiobiological effect. Assuming the dose and the OER do not vary within a voxel, the condition for equal cell survival produces the following equation for the $\mathrm{EUD_{hyp}}$ in hypoxic tumors for a fractionation schedule with n equal fractions:

$$S^n\left(EUD_{hyp}, OER_{eff}\right) = \sum_{i=1}^{I} v_i S_i^k(d_i, OER_i), \tag{4.2}$$

where $\mathrm{OER_{eff}}$ is the effective OER in hypoxic tumor. n is the number of fractions in the $\mathrm{EUD_{hyp}}$ fractionation. v_{i} is the fractional tumor voxel volume. S_{i} is the cell survival fraction in voxel i. and d_{i} and $\mathrm{OER_i}$ are the average values of fractional dose and the OER in the voxel respectively.

Substituting Eq. 1.13 for the LQ cell survival model in Eq. 4.2, we obtain a quadratic equation for the $\mathrm{EUD_{hyp}}$:

$$\alpha_{ox} EUD_{hyp} + \beta_{ox} EUD_{hyp}^2 = -C, \tag{4.3}$$

where the parameter C is given by:

$$C = \frac{1}{n} \ln\left\{\sum_{i=1}^{I} v_i S_i^k(d_i, OER_i)\right\}. \tag{4.4}$$

The standard positive solution to quadratic Eq. 4.3 produces an equation for the $\mathrm{EUD_{hyp}}$:

$$EUD_{hyp} = OER_{eff} \frac{-\alpha_{ax} + \sqrt{\alpha_{ox}^2 - 4\beta_{ox} C}}{2\beta_{ox}}, \tag{4.5}$$

If the parameter β_{ox} in the LQ cell survival model can be omitted, the equation for the EUD_{hyp} can be simplified to:

$$EUD_{hyp} = -OER_{eff}\frac{C}{\alpha_{ox}}. \quad (4.6)$$

We note that for uniform distribution of dose d, the EUD_{hyp} is defined primarily by the hypoxic regions because they define the "cold" spots in radiotherapy response.

The selection of the effective OER_{eff} in Eq. 4.2 requires some assumption of homogenization of the heterogeneous hypoxic tumor. It follows from Eq. 4.2 that different combinations of the effective OER_{eff} and the EUD_{hyp} can produce equal survival of clonogenic cells. Therefore, correct homogenization is a key component in the derivation and application of the equations for EUD_{hyp}.

4.2.2 Tumor-Effective OER

As we mentioned in the previous Section, cell survival in heterogeneous hypoxic tumors depends on both the physical dose EUD_{hyp} and the effective tumor OER_{eff}; therefore, different combinations of the physical dose and the tumor effective OER_{eff} may produce equal cell survival. Therefore, the computation of EUD_{hyp} becomes underdetermined under these conditions and cannot be used for comparison of radiobiological effects in different patients. This problem can be solved by imposing constraints on the OER_{eff}; for example, using homogenization of heterogeneous hypoxic tumors and computing the EUD_{hyp} under the reference oxygenation conditions while maintaining the same radiobiological effect.

It is feasible to select the tumor effective $OER_{eff} = 1$ in Eq. 4.2 corresponding to aerobic conditions for all tumors in the comparison group. Selection of $OER_{eff} = 1$ has many advantages: First, it is a tumor independent constant value for comparison of cell survival in different hypoxic tumors. In this case, all hypoxic tumors with different OER distribution will be compared at the same "reference" aerobic conditions with OER = 1. Second, it provides the condition EUD_{hyp} = EUD for completely oxygenated tumors. Therefore, the effective $OER_{eff} = 1$ is selected for comparison of different treatments plans and tumors.

To follow the standard terminology, we call the EUD_{hyp} with $OER_{eff} = 1$ the Equivalent Uniform Aerobic Dose (EUAD) [18]:

$$EUAD = EUD_{hyp}\left(OER_{eff} = 1\right). \quad (4.7)$$

The EUAD assumes that for any non-uniform distributions of radiation dose delivered to heterogeneous hypoxic tumor and non-uniform distribution of the OER there is a uniform distribution of radiation dose under hypothetical aerobic conditions with $OER_{eff} = 1$ that produces equal survival of clonogenic cells [18]. The EUAD allows

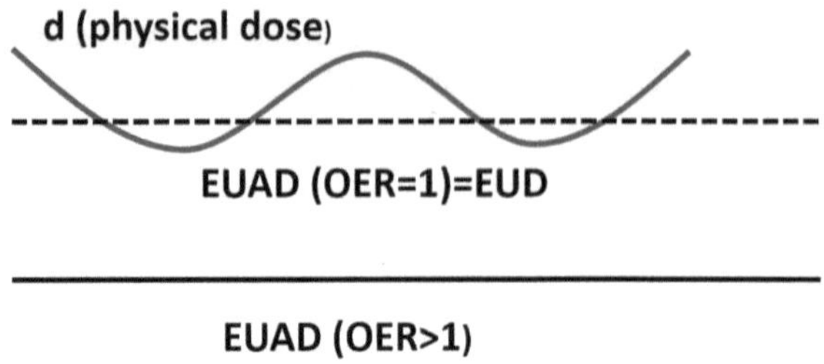

Fig. 4.1 Comparison of EUD and EUAD concepts for a given distribution of physical dose d and different values of OER

one to compare survival of clonogenic cells in tumors with intra-tumor and inter-tumor variation radio sensitivity because cell survival is recomputed under the same benchmark oxygen level ($OER_{eff} = 1$).

The EUAD is a more general concept than the EUD because it is applicable to both oxygenated and hypoxic tumors. The EUAD is equal to the EUD under condition OER = 1 for completely oxygenated tumors. It is usually assumed that the EUD changes between minimum and maximum physical dose. The EUAD changes within much broader range in the following limits:

$$\frac{min\,physicaldose}{maxOER} \leq EUAD \leq \max\ physical\ dose. \tag{4.8}$$

The application of the EUAD and its relationship to the standard EUD is illustrated in Fig. 4.1.

We note that expressing radioresistance of tumor hypoxia in the format of a dosimetry concept such as the EUAD provides physicians with a convenient tool for evaluation of tumor radioresistance because the effective dose may be compared to the prescription dose. The small values of the EUAD would indicate suboptimal treatment plans and high-risk patients. The resulting EUAD can be used for predicting and analyzing treatment outcomes in patients with intra- and inter-tumor variations in radiosensitivity, as well as for biological optimization in radiotherapy treatment planning.

4.2.3 Cellular Proliferation

Cellular proliferation does not have impact on calculation of the EUD because the effect of proliferation depends only on the overall treatment time and does not depend on the radiation dose. Therefore, the proliferation factor is canceled out if the EUD is calculated for equal treatment time [1], The impact of cellular proliferation on the EUAD is different from the EUD because the computation of the EUAD involves hypoxic regions where is usually assumed in calculation that's the hypoxic cells do not proliferate.

If oxygenated cells proliferate, and hypoxic cells do not proliferate, Eq. 4.2 can be re-written as:

$$e^{\lambda T_n} S^n(EUAD) = e^{\lambda T_k} \sum_{i=1}^{Iox} v_i S_i^k(d_i) + \sum_{i=1}^{Ihyp} v_i S_i^k(d_i, OER_i), \tag{4.9}$$

where I_{ox} and I_{hyp} are the number of oxygenated and hypoxic voxels and T_{n} and T_{k} are the total treatment times for the initial and the EUAD dose regimens. Equation 4.9 produces Eq. 4.5 for the EUAD with the parameter C_{p} given by:

$$C_p = \frac{1}{n} \ln \left\{ e^{-\lambda(T_n - T_k)} \sum_{i=1}^{Iox} v_i S_i^k(d_i) + e^{-\lambda T_n} \sum_{i=1}^{Ihyp} v_i S_i^k(d_i, OER_i) \right\}. \tag{4.10}$$

It follows from Eq. 4.10 that cellular proliferation increases the effective value of the EUAD. The impact of cellular proliferation on the EUAD can be also evaluated using the following approximations. The cell survival fraction in oxygenated regions is much smaller compared to hypoxic regions; therefore, Eq. 4.10 can be simplified:

$$C_p \approx \frac{1}{n} \ln \left\{ \sum_{i=1}^{Ihyp} v_i S_i^k(d_i, OER_i) \right\} - \frac{\lambda T_n}{n}. \tag{4.11}$$

Comparing further Eqs. 4.4 and 4.11 we obtain that the correction for cellular proliferation increases the effective value of EUAD.

4.2.4 Absolute Hypoxic Volume

The simulations described in Chap. 3 demonstrate that tumor hypoxia introduces a significant volume effect in the estimates of TCP. We have shown in Chap. 3 that in the presence of a hypoxic region, the TCP is defined entirely by the hypoxic volume. The extent of tumor hypoxia is a plausible mechanism to explain the TCP reduction with increasing total tumor volume observed in clinical studies. Our simulations have also shown that, at the dose levels used in conventional radiation therapy, the TCP dependence on the total tumor volume is negligible for completely oxygenated tumors.

To predict these clinical outcomes, a correction for the absolute hypoxic volume should be implemented in the EUAD to describe the equivalent TCP reduction with the hypoxic volume. This correction should produce smaller values of the EUAD for tumors with larger hypoxic volumes where the TCP reduction is observed. The effect of absolute hypoxic volume can be expressed as an additional volume dependent radioresistance in the TCP. An increase in the hypoxic volume produces a reduction in the TCP. An equivalent reduction in the TCP can be obtained using an increase in the survival fraction leading to a negative correction for the EUAD,

For example, we can assume condition of equal TCP:

$$TCP(V_{hyp}, EUAD) = TCP(V_{hyp,ref}, EUAD - \Delta EUAD_V), \tag{4.12}$$

where $V_{\text{hyp,}}$ is the hypoxic volume and $V_{\text{hyp, ref}}$ is a reference hypoxic volume that can be selected as an average hypoxic volume in the study. Therefore, we obtain the following condition for the reduction ΔEUAD$_V$ in the EUAD with variation of tumor volume:

$$V_{hyp} S^k(EUAD) = V_{hyp,ref} S^k(EUAD - \Delta EUAD_V), \tag{4.13}$$

where S is the cell surviving fraction and k is the number of fractions.

The simulations described in Chap. 3 demonstrate that tumor hypoxia introduces an additional volume dependent radioresistance. The EUAD correction for the absolute hypoxic volume can be derived in the analytical form using Eqs. 4.13 and 1.13 for the LQ cell surviving fraction. Using Eq. 4.13 we obtain:

$$\frac{S(EUAD - \Delta EUAD_V)}{S(EUAD)} = \left\{\frac{V_{hyp}}{V_{hyp,ref}}\right\}^{\frac{1}{k}}. \tag{4.14}$$

Substituting the LQ cell surviving fraction given by Eq. 1.13 in Eq. 4.14 we obtain the following quadratic equation for the correction ΔEUAD$_V$ for the absolute hypoxic volume:

$$-\beta_{ox}\Delta EUAD_V^2 + \Delta EUAD_V(a_{ox} + 2\beta_{ox}EUAD) = \frac{1}{k} ln\left\{\frac{V_{hyp}}{V_{hyp,ref}}\right\} \tag{4.15}$$

Equation 4.15 for corrections ΔEUAD$_V$ can be simplified if we assume $\beta_{ox} = 0$ in the *LQ* model:

$$\Delta EUAD_V = \frac{1}{a_{ox}k} \ln\left\{\frac{V_{hyp}}{V_{hyp,ref}}\right\}. \tag{4.16}$$

The EUAD should be computed using Eq. 4.5 with the subtraction of the correction for absolute hypoxic volume ΔEUAD given by Eqs. 4.15 or 4.16.

4.2.5 EUAD in Radiotherapy Model Problem

In Fig. 4.2, we show a model tumor with a hypoxic region which is irradiated with uniform radiotherapy dose. The hypoxic region may have different OER and volume

V_{hyp}. The dose fractionation schedule is based on total cumulative tumor dose of 60 Gy delivered in 30 fractions. The radiobiological parameters $a_{ox} = 0.28\ \text{Gy}^{-1}$, $\beta_{ox} = 0.045\ \text{Gy}^{-2}$ of the LQ cell survival model for glioblastoma under aerobic conditions have been used in the model problem [19]. The dependence of the EUAD on the OER $\leq$ OER$_{max} = 2$ and the hypoxic volume R $= V_{hyp}/V_{hyp,ref}$ (normalized to a reference hypoxic volume equal to 20% of the total tumor volume) is shown in Fig. 4.3a, b respectively.

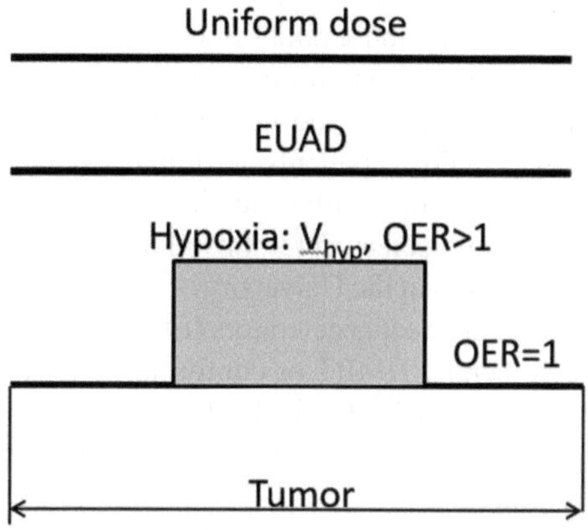

Fig. 4.2 A model tumor with variable OER and hypoxic volume

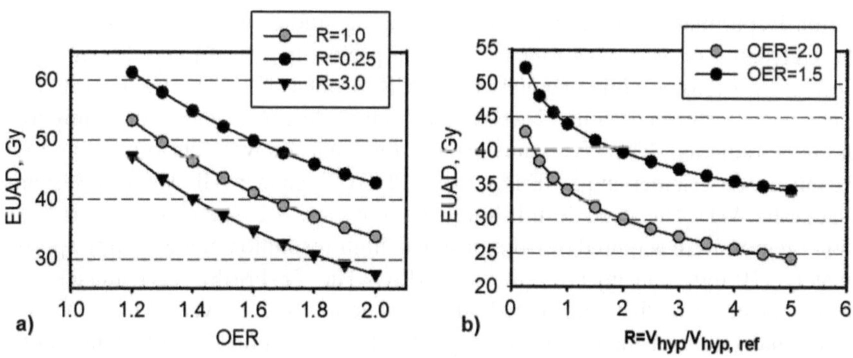

Fig. 4.3 Dependences of the EUAD on OER (a) and hypoxic volume for OER = 2 (a) are shown for the model tumor problem [18]

4.3 Prognostic Value of EUAD

4.3.1 Clinical Data in Radiotherapy for Glioblastoma

To investigate the predictive power of the EUAD, we computed a linear correlation between the OS and the EUAD derived from ^{18}F-FMISO PET hypoxia images in radiotherapy for newly diagnosed glioblastoma multiforme (GBM) (World Health Organization (WHO) grade IV) [18]. The GBM is the most common and most aggressive primary intracranial brain tumor, where the prognosis remains tremendously poor despite modern treatment regimes. Despite treatments with surgery, radiation and chemotherapy, the median OS in patients with glioblastoma is less than 15 months [20]. The GBM patients used in this study were imaged after surgical resection and before radiation with concomitant chemotherapy.

Patients selected for this study have been treated in the Radiation Oncology Department at the University of Washington Medical Center between 2001 and 2016 on Elekta linear accelerators (Elekta, Stockholm, Sweden) using conventional X-ray radiotherapy (IMRT or conformal) with the total radiation dose of 60 Gy in 2 Gy fractions or 61.2 Gy in 1.8 Gy fractions. Treatment planning was performed using in-house software Prism until 2005 and Pinnacle 3D (Philips Healthcare, Fitchburg, WI) later.

Prior to the start of radiotherapy with concomitant chemotherapy, patients underwent ^{18}F-FMISO PET imaging to measure tumor hypoxia in the University of Washington Department of Radiology. These imaging studies were conducted through the University of Washington P01 CA42045 protocols [13, 16] and the American College of Radiology Imaging Network (ACRIN-6684) [21]. The ^{18}F-FMISO PET scans were performed within 2 weeks after tumor diagnosis and prior to initiation of chemoradiation. The PET scans used in this analysis were performed on a GE Advance PET scanner and a GE Discovery PET/CT scanner (GE Medical Systems, Waukesha, WI). The static 20-min single field of view (FOV) emission scans of the tumor region were acquired in 3-dimensional, high-resolution mode starting approximately 110 min after injection of ^{18}F-FMISO. The ^{18}F-FMISO PET images were decay-corrected to the time of radiotracer injection and converted to SUV units by normalization of the emission image to the recorded dose and patient weight. The attenuation correction was done using a 25-min transmission scan for the GE Advance or a low-dose CT scan for Discovery PET/CT scanner.

We identified 11 patients from the P01 CA42045 study and 11 patients from the ACRIN-6684 study and who would be eligible for our proposed research. The eligibility criteria included the availability of pre-treatment ^{18}F-FMISO PET images and completed course of X-ray radiation therapy. The patients who were treated with neutron therapy or X-ray therapy with partial treatment dose were excluded. The data for 22 selected patients include age, sex, radiotherapy dose, OS, tumor volume, absolute hypoxic volume, relative hypoxic volume, and patient-specific values of

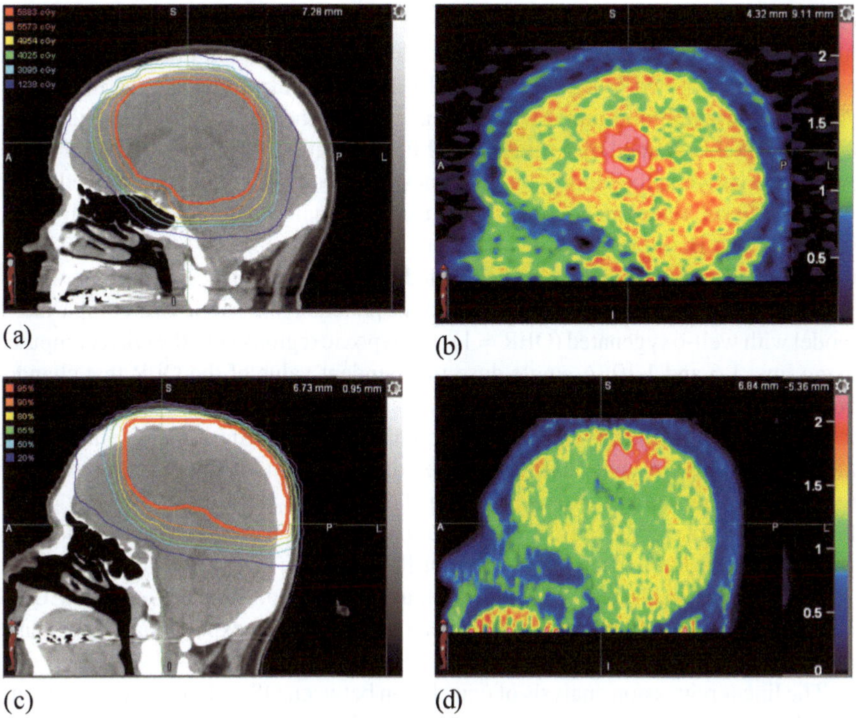

Fig. 4.4 Sample images of dose distribution in radiation therapy computed on CT images (**a**, **c**) and hypoxia in glioblastoma obtained using FMISO PET (**b**, **d**) for two patients

the maximum FMISO uptake in tumors ($SUV_{max,\,p}$), and FMISO uptake in blood ($SUV_{min.\,p}$) [18]. An ^{18}F-FMISO PET image and a radiotherapy dose distribution computed on the CT image for patient are shown in Fig. 4.4 for two sample patients.

4.3.2 Correlation Between EUAD and Overall Survival

The calculation of the EUAD is based on the non-linear mechanistic models of cell survival in hypoxic tumors including the volume effects; therefore, it is expected to provide a better correlation with the tumor control and survival in patients hypoxic tumors compared to the parameters based on imaging features only such as hypoxic volume or tumor/blood ratio. In this article, we study the correlation of EUAD with OS after radiotherapy for glioblastoma.

Analysis of survival data in radiotherapy for hypoxic tumors shows dependence of OS not only on the FMISO SUV values but also on the absolute hypoxic tumor volume. It was shown that the larger hypoxic volumes are associated with poor

OS [13]; therefore, the application of the EUAD requires a correction for the absolute hypoxic volume. The quantification of hypoxic volumes in FMISO studies was previously performed based on venous blood samples drawn during the imaging to calculate a tissue-blood ratio (T/B). In this method, the presence of hypoxia within a tumor voxel is defined by a tumor/blood FMISO ratio that is above a cut off value. As an alternative to the blood samples, a new method based on image-derived blood surrogate tissue reference regions can be used for quantification of FMISO hypoxic volumes [22]. Based on prior data, this cut off value was set to T/B = 1.2 [23] and the hypoxic tumor volumes were delineated using this cut off value.

To compute the EUAD for glioblastoma patients, we use a two-compartment model with well-oxygenated (OER = 1) and hypoxic regions with the OER computed using Eqs. 1.6 and 1.10. A single dose-independent value of the OER that changes between $OER_{min} = 1.0$ and $OER_{max} = 2.0$ is used that is acceptable approximation for radiotherapy with conventional fractionation. Based on the patient data [13]. the following minimum and maximum values of SUV in the patient group have been selected for computing the pO_2 using Eq. 1.10: $SUV_{min} = 1.2$ and $SUV_{max} = 5.0$. For uniform dose prescription, the variation of cell survival in the LQ model is predominantly due to variation in the OER; therefore, nonuniformity in the dose distribution was neglected. We also note that. for uniform dose delivery, the EUAD is primarily defined by the hypoxic regions because of the exponential dependence of radiotherapy response.

The linear regression analysis of correlation between OS and radiotherapy parameters (hypoxic volume and EUAD) was performed using SigmaPlot 12.5 (Grafiti, Palo Alto, CA). The quality of linear regression fit is defined by the linear correlation coefficient squared (R^2) and its significance (*P* value) [24]. The linear correlation between hypoxic volumes and OS for patients is shown in Fig. 4.5 ($R^2 = 0.39$, $P < 0.01$). The linear correlation between the EUAD and OS is shown in Fig. 4.6a. One can see that the EUAD is a predictor of OS with $R^2 = 0.54$ ($P < 0.001$). Figure 4.6b shows that the correction for hypoxic volume improves the linear correlation between the EUAD and OS and the linear correlation coefficient increases to $R^2 = 0.60$ ($P < 0.001$). The

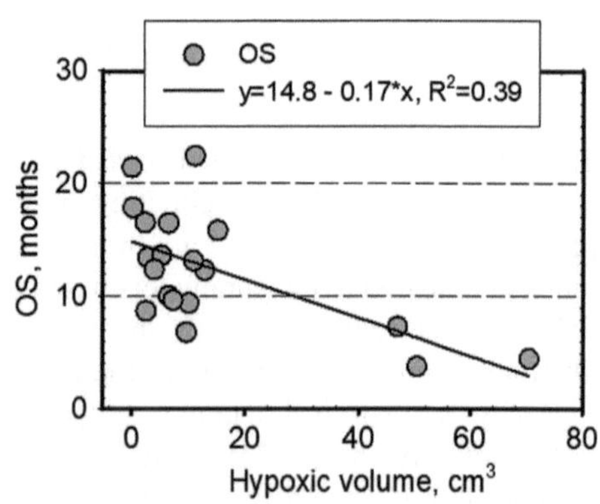

Fig. 4.5 Linear correlation between hypoxic volume and OS in radiotherapy for glioblastoma [18]

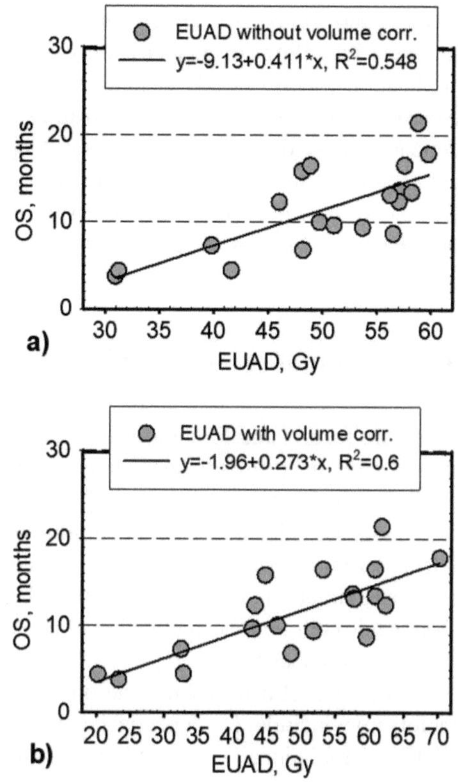

Fig. 4.6 Linear correlation between the EUAD and OS in radiotherapy for glioblastoma: **a** EUAD computed without correction for absolute hypoxic volume; **b** EUAD commuted with correction for absolute hypoxic volume [18]

correlation of the EUAD with OS (with and without correction for hypoxic volume) is better than the correlation of hypoxic volume with OS.

4.4 Application and Limitations of EUAD

4.4.1 Application of EUAD

The EUAD is a universal radiobiological concept that is introduced to extend the concept of EUD and its advantages to heterogeneous hypoxic tumors, The EUAD t is not associated with any specific cancer or any specific MRI or PET biomarkers of hypoxia and can be used in treatment outcomes analysis and optimization for hypoxic tumors. While the analysis proposed in this research is based on the PET FMISO images for glioblastoma; this research can be generalized to other cancers, for example stage III lung cancer, and to other hypoxia biomarkers.

There are several practical applications for the EUAD in radiation therapy for heterogeneous hypoxic tumors. First, the EUAD can be used in treatment outcomes analysis and comparison effectiveness of different treatment plans for different dose regimens including advanced hypofractionated dose regimens. The second practical application for the EUAD in radiation therapy for heterogeneous hypoxic tumors includes the patient-specific maximization of the EUAD using the non-uniform OER-dependent dose distributions known as "dose painting". The larger values of EUAD are associated with improved OS; therefore, it is expected that maximization of the EUAD would improve patient survival. The algorithms developed previously for optimization of the EUD can be applied to optimization of the EUAD [25].

4.4.2 Limitations of EUAD

There are factors and limitations that may affect computing of the EUAD. Some of these factors have already been already considered for the EUD and they can also be applied to the EUAD. For example, the correction for non-uniform spatial distribution of clonogenic cells and the convolution-based correction for inter-patient variation intrinsic radiosensitivity have been considered for the EUD [1]. In this Chapter, our approximations for computing the EUAD included the uniform distribution of clonogenic cells and the application of mean parameters of the LQ model to describe intrinsic radiosensitivity.

It is also useful to discuss the other factors that may affect the EUAD. These factors may include: (1) dependence of the OER on relative biological effectiveness (RBE) of radiation, (2) tumor reoxygenation and acute hypoxia, (3) concomitant chemotherapy.

(1) **tumor reoxygenation and acute hypoxia**

In our applications of the EUAD, we assumed the worst case (most conservative) scenario for the EUAD evaluation in which no re-oxygenation of hypoxic regions of the tumor occurs during treatment. If the serial ^{18}F-FMISO PET images taken during the treatment are available, the EUAD can be re-computed using these images and the corrected value of EUAD will depend on the patient-specific tumor reoxygenation rate. Serial imaging with ^{18}F-FMISO PET for non-small cell lung cancer showed that reoxygenation process was slow (2% day for volumetric re-oxygenation) [26].

Acute or "perfusion" hypoxia, another dynamic process is highly unpredictable and can occur at any time during radiation; therefore, inclusion of acute hypoxia in the analysis of treatment outcomes and radiotherapy optimization is challenging. This would require application of new imaging methods such as Blood Oxygen Level Dependent (BOLD) MRI that can produce quick readouts, allowing investigators to instantly monitor any changes in intravascular pO2 status [27]. Also, this analysis may require statistical methos to describe the random nature of acute hypoxia.

(2) **concomitant chemotherapy**

Concomitant chemotherapy may also have some impact on the EUAD; however, it is expected that this impact is small. Due to large radiation doses, cell killing by radiotherapy dominates in localized primary tumors during concomitant chemotherapy. Additionally, the effectiveness of chemotherapeutic agents in tumors decreases with distance from a capillary because the drug concentration falls off and because of reduced metabolism of hypoxic cells. The drug dose response for many chemotherapeutic agents resembles those for radiation, with drag concentration replacing absorbed dose [28]. In our previous work, we discussed this topic when we reconstructed the cell surviving fractions S_2 (survival at 2 Gy) from imaging data during chemoradiation [29, 30]. The distribution of reconstructed cell surviving fractions agreed with the distributions obtained from in vitro predictive assays.

(3) **dependence of the OER on relative biological effectiveness (RBE)**

The application of these factors and related corrections should be considered based on the specific radiotherapy treatment. The EUAD equations derived in this article assume low-LET radiation with RBE = 1. For high-LET radiation where both OER and RBE may vary, a combination of the EUAD and the equivalent uniform RBE-weighted dose can be used [31], The EUAD can be also considered as equivalent uniform OER-weighted dose.

References

1. Niemierko A. Reporting and analyzing dose distributions: A concept of equivalent uniform dose. Med Phys. 1997; 24: 103–110
2. Ebert M A. Viability of the EUD and TCP concepts as reliable dose indicators. Phys Med Biol. 2000; 45: 441–457.
3. Brahme A. Which parameters of the dose distributions are best related to the radiation response of tumors and normal tissues? In "Radiation dose in radiotherapy from prescription to delivery", Proceedings of the Interregional Seminar for Europe, the Middle east and Africa organized by the IAEA, Leuven; 1994.
4. Mavroidis P, Lind BK, Brahme A. Biologically effective uniform dose (D) for specification, report and comparison of dose response relations and treatments plans. Phys Med Biol. 2001; 37: 2607–2630
5. Yorke ED, Wessels BW, Bradley EW. Absorbed dose averages and dose heterogeneities in radioimmunotherapy. Antibody, Immunoconjugates, and Radiopharmaceuticals. 1991; 4(4): 623–629.
6. Niemierko A. A generalized concept of Equivalent Uniform Dose (EUD). Med Phys. 1999; 28: 1100 (abstract).
7. Wu Q, Mohan R, Niemierko A, Schmidt-Ullrich R. Optimization of intensity-modulated radiotherapy plans based on the equivalent uniform dose. Int J Radiat Oncol Biol Phys. 2002; 52: 224–235.
8. Choi B, Deasy JO. The generalized equivalent uniform dose function as a basis for intensity-modulated treatment planning. Phys Med Biol. 2002; 47: 3579–3589.
9. Jones L. Hoban P. A comparison of physically and radiobiologically based optimization for IMRT. Med Phys. 2002; 29: 1447–1455.

10. Mohan R, Mageras GS, Baldwin B, Brewster LJ, Kutcher GJ, Leibel S, Burman CM, Ling CC, Fuks Z. Clinically relevant optimization of 3-D conformal treatments. Med Phys. 1992; 19: 933–944.
11. Amols HI, Clifton CL. EUD but not QED. Int J Radiat Oncol Biol Phys. 2002; 52: 1–2.
12. Rajendran JG, Schwartz DL, O'Sullivan J, Peterson LM, Ng P, Scharnhorst J, Grierson JR, Krohn KA. Tumor hypoxia imaging with [F-18] fluoromisonidazole positron emission tomography in head and neck cancer. Clin Cancer Res. 2006; 12(18): 5435–41.
13. Spence AM, Muzi M, Swanson KR, O'Sullivan F, Rockhill JK, Rajendran JG, Adamsen TC, Link JM, Swanson PE, Yagle KJ, Rostomily RC, Silbergeld DL, Krohn KA. Regional hypoxia in glioblastoma multiforme quantified with [18F] fluoromisonidazole positron emission tomography before radiotherapy: correlation with time to progression and survival. Clin Cancer Res. 2008; 14:2623–30.
14. Mayr NA, Huang Z, Wang JZ, Lo SS, Fan JM, Grecula JC, Sammet S, Sammet CL, Jia G, Zhang J, Knopp MV, Yuh WT. Characterizing tumor heterogeneity with functional imaging and quantifying high-risk tumor volume for earlier prediction of treatment outcome: cervical cancer as a model. Int J Radiat Oncol Biol Phys. 2012; 83(3):972–979.
15. Bowen SR, Yuh WTC, Hippe DS, Wu W, Partridge SC, Elias S, Jia G, Huang Z, Sandison GA, Nelson D, Knopp MV, Lo SS, Kinahan PE, Mayr NA. Tumor radiomic heterogeneity: Multiparametric functional imaging to characterize variability and predict response following cervical cancer radiation therapy. J Magn Reson Imaging. 2018; 47(5):1388–1396.
16. Muzi M, Wolsztynski E, Fink JR, O'Sullivan JN, O'Sullivan F, Krohn KA, Mankoff DA. Assessment of the Prognostic Value of Radiomic Features in ^{18}F-FMISO PET Imaging of Hypoxia in Postsurgery Brain Cancer Patients: Secondary Analysis of Imaging Data from a Single-Center Study and the Multicenter ACRIN 6684 Trial. Tomography. 2020; 6(1): 14–22.
17. Chen G-P, Ahunbay E, Schultz C, Allen Li X. Development of an inverse optimization package to plan nonuniform dose distributions based on spatially inhomogeneous radiosensitivity extracted from biological images. Med. Phys. 2007, 34: 1198–1205.
18. Chvetsov AV, Muzi M. Equivalent uniform aerobic dose in radiotherapy for hypoxic tumors. Phys. Med. Biol. 2024; 69: 085011.
19. Wigg DR. Applied Radiobiology and Bioeffect Planning. Madison, WI. Medical Physics Publishing; 2001.
20. Stupp R et al. Radiotherapy plus concomitant and adjuvant temozolomide for glioblastoma. N Engl J Med. 2005; 352: 987–996.
21. Gerstner ER, Zhang Z, Muzi M, Hanna L, Greco E, Prah M, Schmainda KM, Mintz A, Kostakoglu L, Eikman EA, Ellingson BM, Ratai EM, Sorensen AG, Barboriak DP, Mankoff DA. ACRIN 6684: Assessment of Tumor Hypoxia in Newly Diagnosed Glioblastoma Using 18F-FMISO PET and MRI. Clin Cancer Res. 2016; 22:5079–5086.
22. Muzi M, Peterson LM, O'Sullivan JN, Fink JR, Rajendran JG, McLaughlin LJ, Muzi J P, Mankoff DA, Krohn KA. 18F-Fluoromisonidazole Quantification of Hypoxia in Human Cancer Patients Using Image-Derived Blood Surrogate Tissue Reference Regions. J. Nucl. Med. 2015; 56: 1223–8.
23. Rasey JS, Koh WJ, Evans ML, Peterson LM, Lewellen TK, Graham MM, Krohn KA. Quantifying regional hypoxia in human tumors with positron emission tomography of [^{18}F] fluoromisonidazole: A pretherapy study of 37 patients. Int J Radiat Oncol Biol Phys. 1996; 36(2): 417–428.
24. Nickoloff E L 2011 *Applications of Statistics to Medicine and Medical Physics*. (Madison, Wis., Medical Physics Publishing).
25. Chvetsov AV, Dempsey JF, Palta JR. Optimization of equivalent uniform dose using the L-curve criterion. Phys Med Biol. 2007; 52: 5973–5984.
26. Koh WJ, Bergman KS, Rasey JS, Peterson LM, Evans ML, Graham MM, Grierson JR, Lindsley KL, Lewellen TK, Krohn KA, Griffin TW. Evaluation of oxygenation status during fractionated radiotherapy in human non-small cell lung cancer using (F-18) fluromisonidazole positron emission tomography. Int J Radiat Oncol Biol Phys. 1995; 33(2): 391–398.

27. Tatum JL, Kelloff GJ, Gillies RJ, Arbeit JM, Brown JM, Chao KS, Chapman JD, Eckelman WC, Fyles AW, Giaccia AJ, Hill RP, Koch CJ, Krishna MC, Krohn KA, Lewis JS, Mason RP, Melillo G, Padhani AR, Powis G, Rajendran JG, Reba R, Robinson SP, Semenza GL, Swartz HM, Vaupel P, Yang D, Croft B, Hoffman J, Liu G, Stone H, Sullivan D. Hypoxia: importance in tumor biology, noninvasive measurement by imaging, and value of its measurement in the management of cancer therapy. Int J Radiat Biol. 2006; 82(10):699–757.
28. Hall EJ, Giaccia AJ. Radiobiology for the Radiologist. 6th ed. Philadelphia, PA: Lippincott Williams & Wilkins; 2006.
29. Chvetsov AV, Yartsev S, Schwartz JL, Mayr N. Assessment of interpatient heterogeneity in tumor radiosensitivity for non-small cell lung cancer using tumor-volume variation data. Med. Phys. 2014; 41(6): 064101-1–10.
30. Chvetsov AV, Sandison GA, Schwartz JL, Rengan R. Ill-posed problem and regularization in reconstruction of radiobiological parameters from serial tumor imaging data. Phys Med Biol. 2015; 60(21): 8491–8503.
31. Chvetsov AV. Equivalent uniform RBE-weighted dose in eye plaque brachytherapy Med. Phys. 2024; 51(4): 3093–3100.

Chapter 5
Dose Nonuniformity Effectiveness

5.1 Biological Effectiveness of Nonuniform Dose Distributions

5.1.1 Hypoxia-Targeted Dose Distributions

It has been shown in clinical trials with the PET and MRI hypoxia imaging that the increased tumor hypoxic fractions deteriorate of the TCP and overall survival [1–3]. It is believed that the PET and MRI images of tumor hypoxia can be used for the development of radiotherapy techniques utilizing non-uniform dose distributions targeting radioresistant hypoxic regions with higher doses to improve the TCP. The effectiveness of non-uniform hypoxia-targeted dose distributions in heterogeneous hypoxic tumors is supported by the theoretical computational studies based on the performance criteria utilizing the analytical tumor control probability (TCP) equation or integrated cell survival [4–7]. The uniform and non-uniform dose distributions for tumors with heterogeneous radiosensitivity have been compared under constraints of integral tumor dose. It was shown in these studies that the non-uniform hypoxia targeted dose distributions may improve tumor-integrated cell killing compared to uniformly escalated dose distributions. The hypoxia-targeted dose escalation has been investigated in conventionally fractionated radiotherapy and hypofractionated radiotherapy [8–13].

In the current practical radiation oncology, physicians rely mostly on re-oxygenation to overcome the hypoxia-induced radioresistance. However, the observations that pre-therapy oxygen status is correlated with radiotherapy outcomes cast doubt on the effectiveness of re-oxygenation in human tumors [2, 14]. Escalating dose in radioresistant hypoxic regions can promote tumor cell death and potentially improve treatment outcomes. Conversely, patients with well-oxygenated tumors may benefit from dose de-escalation allowing for better normal tissue protection [15]. An effective dose modulation technique that improves the treatment outcomes would

A. V. Chvetsov, *Optimization Problems in Radiotherapy for Hypoxic Tumors*, Biological and Medical Physics, Biomedical Engineering,
https://doi.org/10.1007/978-981-96-7001-7_5

have a significant clinical impact on many patients. Therefore, treatment techniques such as hypoxic dose escalation which may help battle hypoxia should be further explored.

Challenges for the clinical application of hypoxia targeted dose escalation include the risks associated with normal tissue complications and dynamic nature of the location and extent of hypoxia. For example, temporal variation of hypoxic regions, target deformation and tumor motion can generate delivery errors that would compromise precise targeting of hypoxia using non-uniform dose distribution. Many research studies on functional image-guided dose escalation have been treatment planning studies, and treatment outcome have not been studied. Therefore, hypoxia-targeted dose escalation has not yet found a widespread application in the clinic. More robust alternative approaches of targeting hypoxia with radiation are needed that are less sensitive to these uncertainties. Apparently, the most robust dose escalation would be uniform to the entire tumor volume because it is not influenced by the location of the hypoxic sub-volume.

The feasibility of hypoxia targeting dose escalation in radiotherapy is under investigation, however some radiotherapy techniques, for example stereotactic body radiation therapy (SBRT), already deliver nonhomogeneous dose distributions that may counteract hypoxia [16]. Dose escalation to the central part of the tumor is a part of standard treatment in SBRT because the dose is prescribed to 60–90% isodose line. This creates a non-uniform dose distribution with higher doses at the center of the tumor and lower doses at the periphery (equal to prescribed dose). While the feasibility of hypoxic boost in conventional radiotherapy is under discussion, this integrated dose escalation in SBRT may potentially deliver higher dose to the hypoxic region [16]. The studies with the TCP modelling [17] and single fraction simulations of relative cell survival [5, 6] determined that the hypoxia-targeted dose escalation is more efficient in hypofractionated dose regimens than in conventional fractionation.

5.1.2 *Effects of Dose Uniformity and Nonuniformity*

As we already mentioned, the effectiveness of non-uniform hypoxia-targeted dose distributions in heterogeneous hypoxic tumors is supported by the theoretical computational studies based on the performance criteria utilizing the analytical tumor control probability (TCP) equation or integrated cell survival, However, some studies show that a uniform dose escalation may be more efficient than hypoxia-targeted dose escalation when both distributions maintain an equal integral tumor dose. For example, it was shown demonstrated that the uniform dose escalation could produce a higher cell killing effect than the dose boosting to hypoxic regions in radiotherapy with conventional fractionation [5, 6]. While apparently being limited to a single dose fraction, these theoretical results nevertheless seem to argue against dose painting as an efficient treatment strategy under certain conditions.

It can be also theoretically shown that the uniform dose distributions maximize the TCP in radiotherapy for tumors with uniform distribution of radiosensitivity assuming equal integral tumor dose [18]. Based on this theoretical result, the nonuniform dose distributions are always less efficient in homogeneous tumors. Assuming continuous variation of effectiveness, it is feasible to suggest that increasing tumor heterogeneity may create a transitional zone where the uniform dose distributions remain more efficient for relatively small variation of radiosensitivity. The analysis of hypoxia images in clinical trials shows that that many human tumors have the maximum value of partial oxygen pressure (pO_2) below the level of hypoxia leading to smaller variation of radioresistance [7]. In this Chapter, we perform a theoretical analysis of biological effectiveness of non-uniform pO_2-targeted dose distributions for tumors with intermediate oxygen levels to understand if the targeting these regions with higher doses is efficient.

When developing these "dose painting" protocols, it is useful to determine quantitative criteria that define effectiveness of these protocols and evaluate the factors that impact the optimal dose escalation. The final effectiveness of dose escalation protocols can be determined only from clinical trials; however, some criteria of effectiveness can be derived from basic radiobiological equations that can provide guidelines for these clinical trials. The concepts of TCP, EUAD and tumor response models considered in Chaps. 2–4 can be used to understand uncertainties and improve biological planning for radiation therapy for hypoxic tumors.

5.1.3 Dose Nonuniformity Effectiveness

In the development of the OER-modulated dose escalation protocols, it is important to determine whether a non-uniform hypoxia targeted dose escalation produces a substantial increase in cell killing relative to a uniform dose escalation when both distributions are compared under constraints of the equal integral dose to normal tissues. The simplest universal constraint in this comparison may be equal integral dose delivered to the tumor. The non-uniform hypoxia-targeted dose distributions can outperform the uniform dose distributions only if they can produce a substantially higher cell killing effect under condition of equal integral dose delivered to the tumor.

A uniform dose distribution, while technically not achievable in some radiotherapy techniques such as brachytherapy and stereotactic radiosurgery, has many practical advantages in conventional radiotherapy. First, the uniform dose distributions facilitate dose delivery as well as quality assurance and are more suitable for standardization, reporting, and, therefore, outcomes analysis [19]. Second, the uniform dose distributions have an advantage of being unaffected by the location or shape of the hypoxic region that has the tendency to change during treatment due to deformation, motion, or re-oxygenation of gross tumor volume.

For comparison of the biological effectiveness of the uniform and non-uniform dose distributions in heterogeneous hypoxic tumors, we introduce a new parameter called the Dose Nonuniformity Efficiency (DNE) that is based on the concept of

EUAD [14]. To measure the effectiveness of a non-uniform dose distribution, we consider a ratio of the EUAD for a non-uniform and the uniform dose distributions with equal integral tumor dose [7]. We define the DNE as

$$DNE = \frac{EUAD(\mathbf{D}_N)}{EUAD(\mathbf{D}_U)}, \tag{5.1}$$

where EUAD($\mathbf{D}_\mathrm{N}$) is computed for a non-uniform distribution and EUAD($\mathbf{D}_\mathrm{U}$) is the reference value is computed for the uniform dose distribution.

It is obvious that larger values of average physical tumor dose would produce larger values of EUAD; therefore, to study the effect of dose redistribution between radio-resistant and radio-sensitive tumor regions, it is feasible to normalize the EUAD to a reference value obtained for the uniform distribution with equal average dose. Therefore, we assume that both EUAD_N and EUAD_U computed for the equal average tumor dose. We note that the approach of DNE allows for comparison of not only uniform and non-uniform dose distributions but also two non-uniform dose distributions.

Based on the previous studies on biological effectiveness of nonuniform dose distributions in heterogeneous hypoxic tumors, it is expected that DNE > 1 in heterogeneous hypoxic tumors. However, the following condition based on the theory of EUD is also valid:

$$\lim_{\mathrm{OER}\to 1} DNE = \frac{EUD}{D_{av}} \leq 1, \tag{5.2}$$

where D_av is the average tumor dose. The condition described by Eq. (5.2) presents a challenge for hypoxia-targeted nonuniform dose distributions. This condition indicates that the nonuniform dose distributions may lose their advantages compared to uniform dose distributions for the small values of OER or in the case of quick reoxygenation.

In this Chapter, we use the concept of DNE for a detailed comparison of the cell killing effect for non-uniformly and uniformly escalated dose distributions in tumors with heterogeneous radiosensitivity during fractionated radiotherapy using analytical and computational approaches. We compute the DNE for different values of OER, hypoxic tumor volumes, and determine where the non-uniform dose distributions may be feasible. The DNE is analyzed for conventionally fractionated radiotherapy with different initial hypoxic conditions, dose boost ratios, re-oxygenation rates and delivery errors. It is also important to investigate how the DNE changes with increasing fractional dose; therefore, we extended our analysis to include hypofractionated regimens. As an example, we study the hypofractionated dose regimens used in SBRT for non-small cell lung cancer.

5.2 DNE in Model Problems with Dose Escalation

5.2.1 *Two-Level Model for DNE*

(a) **Equations of DNE model**

Our analysis of DNE is done for fractionated radiotherapy using a tumor response model that uses as few parameters as possible to describe the major components of tumor response to irradiation such as radiosensitivity, proliferation and hypoxia. To achieve these goals, we employ the 2-level cell population tumor response model for the TCP analysis described in Chap. 2 that is given by Eqs. 2.14 and 2.15. This tumor model represents the 4-level cell population tumor response model without lethally damaged cells. This model describes major components of tumor response to radiotherapy such as radiosensitivity, proliferation and hypoxia using as few variables as possible. Each of these parameters can be estimated using predictive assays (parameters of the LQ model and a proliferation constant) or functional imaging (hypoxic fraction and re-oxygenation rate). This model also allows us to obtain analytical solutions and other advantages of this model include possible evaluation of model parameters using imaging studies and predictive assays and as well as application to combined radiotherapy and chemotherapy.

We also introduce a radiotherapy model problem with radiobiological parameters related to non-small cell lung cancer is used to study the DNE in radiotherapy with uniform and non-uniform hypoxia-targeted dose escalation that provide equal integral total tumor dose (Fig. 5.1) [6]. We assume that V_{ox} and V_{hyp} are the oxygenated and hypoxic tumor volumes irradiated with doses d and $d_b = bd$ respectively, where $b \geq 1$ is the dose escalation coefficient. To facilitate computation of uniform dose d_U with equal integral dose and link the tumor geometry to the 2-level cell population tumor model given by Eqs. 2.14 and 2.15, we assume that the initial distribution of viable clonogenic cells in the tumor is uniform.

$$n_{m,0} = \rho V_{m,0}, m = ox, hyp, \tag{5.3}$$

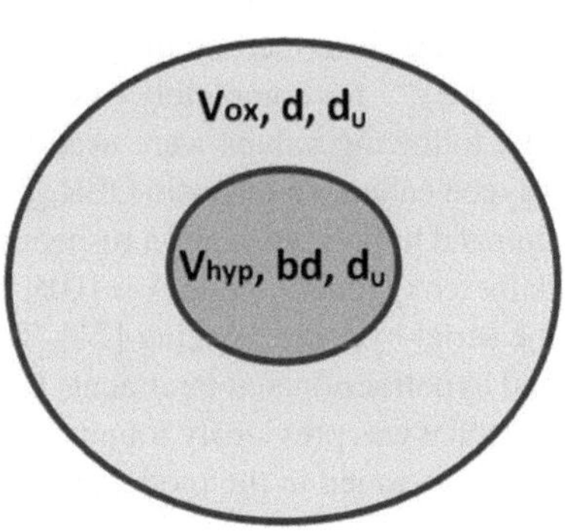

Fig. 5.1 Model radiotherapy problem with tumor dose d, hypoxia-targeted escalated dose bd and uniformly escalated dose d_u [7]

where ρ is the density of clonogens. Using these model parameters, a uniform dose that would produce equal integral dose for non-uniform dose distribution in the tumor can be calculated as:

$$d_U = d\big(1 + v_{hyp}(b - 1)\big), \tag{5.4}$$

where v_{hyp} the fractional hypoxic tumor volume

We use the concept of DNE for detailed analytical and computational comparison of the cell killing effect for non-uniformly and uniformly escalated dose distributions in tumors with heterogeneous radiosensitivity during fractionated radiotherapy as shown in Fig. 5.1. Assuming the tumor model problem with OER dependent dose and the 2-level tumor response model, the DNE at the k-th dose fraction is given by the following equation:

$$DNE_k = \frac{EUAD_k(d, bd)}{EUAD_k(d_U, d_U)}, \tag{5.5}$$

where the EUAD for the uniform and non-uniform dose distributions is, in a general case, is given by Eq. 4.5 with the parameter C computed using the following equation:

$$C(d, bd) = -\frac{1}{k}\ln\big\{v_{ox}S^k(d) + v_{hyp}S^k(bd, OER)\big\} \tag{5.6}$$

or

$$C(d_U, d_U) = -\frac{1}{k}\ln\big\{v_{ox}S^k(d_U) + v_{hyp}S^k(d_U, OER)\big\}, \tag{5.7}$$

where $v_{ox} = V_{ox,0}/V_{tot,0}$ and $v_{hyp} = V_{hyp,0}/V_{tot,0}$ are the initial fractional volumes of hypoxic and oxygenated tumor regions.

(b) **Parameters of the DNE model**

The fractionated dose regimens in the model problem include a conventional fractionation regimen with a total dose of 60 Gy delivered in 30 fractions and a hypofractionated regimen with a total dose of 48 Gy delivered in 4 fractions. The cell survival is described by the LQ-model with the parameters $\alpha = 0.45$ Gy^{-1} and $\beta = 0.039$ Gy^{-2} for completely oxygenated cells which are typical for lung cancer [20]. The following values were used for other computational parameters of the model: oxygen enhancement ratio $\text{OER}_{max} = 2$ and $\text{OER}_{max} = 3$ for conventional fractionation and hypofractionation respectively, potential doubling time $T_{pot} = 7.0$ [20] and daily reoxygenation rate $A = 0.05$. The daily reoxygenation rate was evaluated using the serial hypoxia imaging [21]. The reoxygenation rate was equal in conventional and hypofractionated treatments because we believe it is related to clearance of dead cells that was previously found to be identical in these types of radiotherapy [22, 23]. It was assumed in the model problem that the hypoxic fractional volume $v_{hyp} = 0.2$

and the hypoxic tumor fraction receives an escalated dose characterized by the dose boost ratio b = 1.3.

The initial hypoxic volume is used to calculate the initial boost volume in our simulations; this volume did not change during radiotherapy treatment. In this approach, the re-oxygenated cells receive an escalated boost dose. This problem can be solved using an adaptive boost model where the boost parameters are recalculated based on the reduced hypoxic volume. For mathematical brevity, the boost adaptation is not considered.

5.2.2 Analytical Solutions for DNE

The DNE defined by Eq. 5.5 can be evaluated analytically for some special cases. It is possible to show that DNE ≤ 1 is for some specific conditions, for example, $k = 1$ that means the first dose fraction in conventional radiotherapy and $\beta_{ox} = 0$. If the parameter β_{ox} in the LQ model ca be omitted, the DNE_1 for $k = 1$ can be written as:

$$DNE_1 = \frac{\ln\left\{v_{ox}\exp(-\alpha_{ox}d) + v_{hyp}\exp\left(-\frac{\alpha_{ox}bd}{OER}\right)\right\}}{\ln\left\{v_{ox}\exp(-\alpha_{ox}d_U) + v_{hyp}\exp\left(-\frac{\alpha_{ox}d_U}{OER}\right)\right\}}. \tag{5.8}$$

We present an analytical proof that the DNE for the first dose fraction in conventionally fractionated radiotherapy is always DNE ≤ 1 for the small values of $\alpha_{ox}d$. Using the linear expansion of exponent in Eq. 5.8:

$$\exp(x) \sim 1 - x + O\left(x^2\right), \tag{5.9}$$

that is acceptable approximation for the small values of $\alpha_{ox}d$ and additionally requesting DNE ≤ 1 we obtain the inequality:

$$d\left\{v_{ox} + \frac{v_{hyp}b}{OER}\right\} \le d_U\left\{v_{ox} + \frac{v_{hyp}}{OER}\right\}. \tag{5.10}$$

Using Eq. 5.4 for the average dose d_U in Eq, 5.10 we obtain:

$$\left\{v_{ox} + v_{hyp}b\right\} \le \left\{1 + v_{hyp}(b-1)\right\}\left\{v_{ox} + \frac{v_{hyp}}{OER}\right\}. \tag{5.11}$$

Equation 5.11 can be simplified to

$$\frac{b}{OER} + 1 \le \frac{1}{OER} + b, \tag{5.12}$$

that can be written as

$$(b-1)(OER-1) \geq 0. \tag{5.13}$$

Equation 5.13 is always valid because of initial conditions $b \geq 1$ and $OER \geq 1$; therefore, the initial condition DNE ≤ 1 is also valid. That means that the DNE is always DNE ≤ 1 and uniform dose distribution produces higher larger killing than non-uniform dose distribution under these specific conditions.

The linear expansion of the exponent that was used in the derivation of Eq. 5.13 has an approximation error $\leq 10\%$ when $\alpha_{ox}d \leq 0.4$. There are therapeutically relevant cases that would satisfy the condition $\alpha_{ox}D \leq 0.4$, for example, hyperfractionated dose regimens (two dose fractions a day) for head and neck cancers treated with the fractional dose of 1.2 Gy and $\alpha_{ox} = 0.3\ Gy^{-1}$ [20]. Another clinical case where this condition works is the small cell lung cancer treated with conventional fractionation 2 Gy per fraction and $\alpha_{ox} < 0.2\ Gy^{-1}$ [20]. For clinical cases where $\alpha_{ox}d > 0.4$, we perform computational simulations below.

5.2.3 Computational Solutions for DNE

For most radiotherapy scenarios in the model problem shown in Fig. 5.1, the DNE can be investigated using only computational methods. We show the computed DNE for the uniform dose escalation and non-uniform hypoxia-targeted dose escalation in fractionated radiation therapy with conventional and hypofractionated dose regimens. We assume that both uniform and non-uniform dose distributions have an equal integral tumor dose.

In Fig. 5.2, the DNE is computed for different dose regimens during entire treatment with different reoxygenation rates. In Fig. 5.2a, we show the DNE during conventional radiotherapy. One can see that the DNE varies from values below 1.0 to values above 1.0 during conventional radiotherapy. It agrees with the analytical calculation for the DNE that predicted that the DNE is smaller than 1.0 for the first fraction. One can see that the DNE is usually larger than 1.0 that means that non-uniform dose distribution produces a higher cell killing effect than the uniform dose distribution for most cases. In Fig. 5.2b we show the DNE for hypofractionated regimens that is larger than for conventional fractionation indicating higher effectiveness of hypoxia targeted dose escalation.

The obtained data suggests that while a non-uniform hypoxia targeted dose distribution can be less effective in the conventionally fractionated radiotherapy than the uniform dose distribution at the beginning of it will outperform uniform dose distribution by the end of treatment. For the first few fractions in conventional radiation therapy where initial distribution of clonogenic cell can be considered nearly uniform, our analytical and computational results agree with the single-fraction simulation where it was demonstrated that a uniform dose distribution produces a larger cell killing effect than a non-uniform dose distribution [14]. This effect may arise from the following computational procedures. Computing an average dose for hypoxia-targeted dose escalation produces an increase of integral dose in the oxygenated

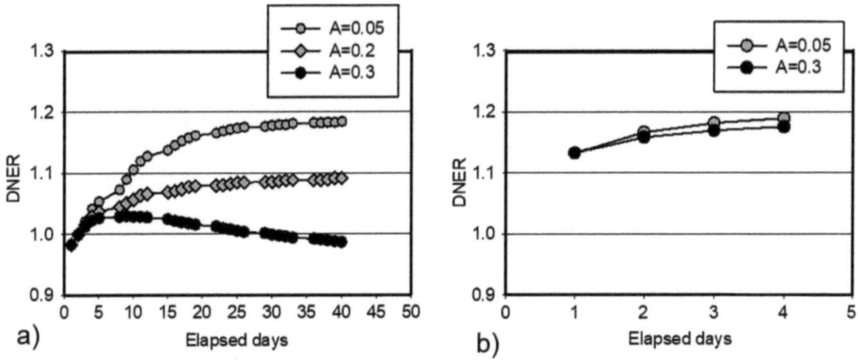

Fig. 5.2 The DNE computed different tumor reoxygenation rates for **a** conventionally fractionated radiotherapy and **b** hypofractionated radiotherapy [7]

tumor fraction and an equivalent decrease of integral dose in the hypoxic region. However, the domination of linear contribution to cell killing in the dose interval of 1–2 Gy and a steeper slope of survival curve in the oxygenated region create an increase of cell killing which is larger than a decrease of cell killing in the hypoxic region. As a result, the average dose produces larger cell killing for an equal integral dose.

For hypofractionated dose regimens, the non-uniform dose distribution always produced DNE > 1 for the entire treatment that mean larger cell killing effect compared to uniform dose distribution. Apparently, this results from domination of quadratic contribution to the cell killing effect in the LQ model for larger doses that counteracts the slower slope of survival curve in the hypoxic region. Toward the end of the treatment, the DNE produced by non-uniform dose distributions is almost the same for hypofractionated regimens and conventional fractionated regimens assuming that the dose escalation factor is $b = 1.3$.

5.2.4 Reoxygenation and Delivery Errors

Tumor reoxygenation and delivery errors may have impact on the DNE for non-uniform hypoxia-targeted dose distributions and should also be investigated. These effects are more important for conventional fractionation because of longer treatment times. The effect of re-oxygenation results in irradiation of re-oxygenated cells with the escalated dose when the escalation volume is permanent. Under delivery errors, we assume the irradiation of a fractional hypoxic volume with non-escalated dose and the irradiation of an equivalent absolute volume in the oxygenated tumor region with the escalated dose. The delivery errors may result from a displacement of the hypoxic region relative to the radiation field.

In Fig. 5.2, the DNE for conventional fractionation and hypofractionation is computed for different reoxygenation rates. One can see that the impact of reoxygenation on the DNE is larger for conventional fractionation. When the reoxygenation rates increase, the DNE approaches value close to 1.0 thus demonstrating the increased efficiency of uniform dose distributions. The delivery errors can be measured as a percentage of the hypoxic region receiving non-escalated dose. While the irradiation pattern is different for reoxygenation and delivery errors, the effect on relative cell survival is very similar.

The non-uniform distribution outperforms uniform distribution towards the end of the treatment where the initial distribution of clonogenic cells for each fraction becomes non-uniform. Reoxygenation that makes tumor more homogeneous is a competing process that could reduce effectiveness of non-uniform dose distribution at the end of the treatment. Delivery errors can lead to irradiation of the hypoxic tumor region with a lower dose, and also reduce effectiveness of non-uniform dose distributions. In general, based on our simulations, the hypoxia-targeted dose distributions in conventional radiotherapy could be more effective than the uniformly escalated dose distributions; however, the domination of linear contribution to cell killing, reoxygenation and delivery errors can reduce the potential of non-uniform dose escalation. We note that a reduction in TCP in the presence of delivery errors for non-uniform hypoxia targeted dose distributions and an increase in TCP in the presence of re-oxygenation for uniform dose distributions was also observed in [4]. The quadratic contribution to the cell killing effect and the reduced negative impact of reoxygenation and delivery errors make the non-uniform hypoxia targeted dose distributions in SBRT very efficient.

5.2.5 DNE for Intermediate Oxygen Levels

In Fig. 5.3a, b, the DNE was computed as a function of OER, dose escalation factor b for conventional fractionation and reoxygenation rate A for conventional dose regimen. Our simulations of DNE show that the entire region of OER in conventionally fractionated and hypofractionated radiotherapy for heterogeneous tumors can be separated in two regions by a threshold value of OER_{th}:

(1) $OER > OER_{th}$ where $DNE > 1$ indicating higher effectiveness of nonuniform dose distributions and
(2) $OER < OER_{th}$ with $DNE < 1$ indicating higher effectiveness of uniform dose distributions where non-uniform targeted dose distributions are not beneficial.

Our simulations show that the value of the threshold OER_{th} in conventional fractionation is the range of about 1.2–1.6 depending on selected radiotherapy parameters. In general, the OER_{th} increases with reoxygenation rate, relative hypoxic volume and dose escalation factor.

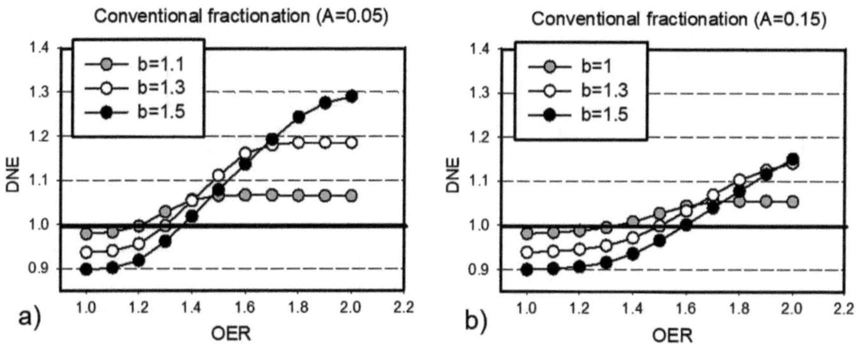

Fig. 5.3 Dependence of the DNE on the OER. dose escalation factor b. and reoxygenation rate A for conventionally fractionated radiotherapy [7]

In Fig. 5.4, we show the dependence of DNE on the OER and dose escalation factor b for hypofractionation. The dependence on reoxygenation for hypofractionation is negligible and is not shown. The threshold value of OER_{th} is about 1.1 for hypofractionated radiotherapy in that is smaller compared to conventional radiotherapy.

These computational results are supported by the theoretical proof that uniform dose distributions produce the largest TCP in tumors with homogeneous radiosensitivity. Comparing uniform and nonuniform dose distributions in a homogeneous tumor, we obtain DNE < 1. The DNE is a continuous function; therefore, slowly increasing tumor heterogeneity, we obtain the condition DNE = 1 only for some threshold value of tumor heterogeneity.

Figure 5.5 shows dependence of the DNE at the end of treatment on different parameters of radiation therapy such as reoxygenation rate A and the dose escalation factor *b*. The variation of DNE for different values of b and A is shown for radiotherapy treatment with conventional fractionation (60 Gy in 30 fractions) and for a

Fig. 5.4 Dependence of the DNE on the OER and dose escalation factor b for hypofractionated radiotherapy [7]

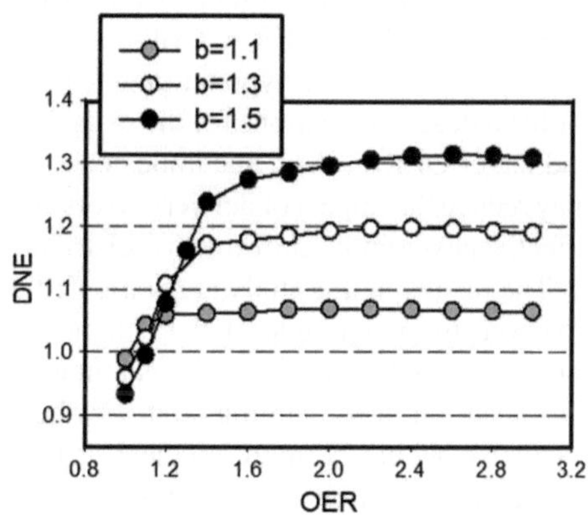

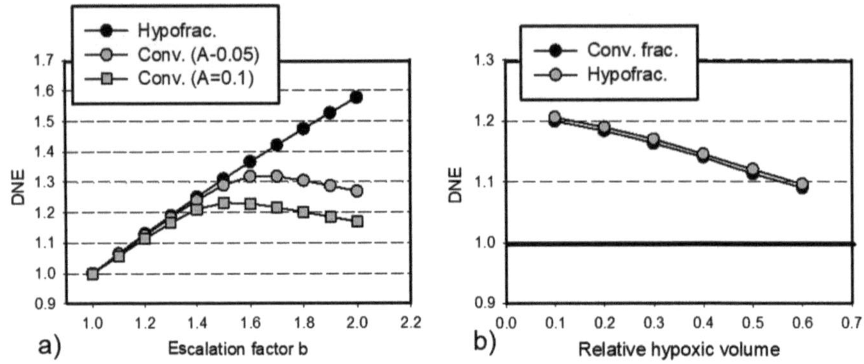

Fig. 5.5 Dependence of the DNE on dose escalation factor b for conventionally fractionated radiotherapy ($OER_{max} = 2$) and hypofractionated radiotherapy ($OER_{max} = 3$) [7]

hypofractionated regimen (48 Gy in 4 fractions). One can see that the DNE increases with the dose escalation factor b and decreases with relative hypoxic volume.

5.2.6 Conclusions

Our simulations show that the entire range of OER can be separated by a threshold value of OER_{th}: (1) OER > OER_{th} where DNE > 1 indicating higher effectiveness of nonuniform dose distributions and (2) OER < OER_{th} with DNE < 1 where non-uniform targeted dose distributions are not beneficial. The most robust dose escalation would be uniform to the entire tumor volume because it is not influenced by the location of the hypoxic sub-volume. Simple radiobiological simulations show that uniform dose escalation can be more efficient in cell killing in the presence of reoxygenation and delivery errors [6, 7]. Therefore, the non-uniform dose distributions would be more efficient for OER > OER_{th} and uniform dose distributions should be used for OER < OER_{th}. We also note that, in hypofractionated radiotherapy, non-uniform hypoxia-targeted dose escalation can be more efficient than uniform dose distributions for the extended range of OER because of smaller OER_{th} and larger OER_{max}. The quadratic cell killing effect and short treatment time which may reduce the impact of delivery errors and re-oxygenation may contribute to higher effectiveness of non-uniform dose distributions in hypofractionated radiotherapy [6].

The results in this study are computed using a simplified tumor response model; however, the magnitude of the obtained effects suggests that these results effects more likely will be observed in real tumors. The results of our simulations may have an impact on dose painting strategies because they suggest that the uniform dose escalation would be more efficient at intermediate oxygen levels. Delivery errors and homogenization of tumor resulting from re-oxygenation should be considered because these factors could reduce the effectiveness of hypoxia-targeted dose escalation in conventional radiotherapy. The reoxygenation and delivery errors are less important for hypoxia targeted dose escalation in hypofractionated dose regimens such as SBRT. The lower impact of reoxygenation and delivery errors combined with quadratic contribution to the cell killing effect can make hypoxia targeted dose escalation very effective in such hypofractionated dose regimens.

5.3 Volume Effects for Hypoxia-Targeted Dose Distributions

5.3.1 Limited Dose Escalation

The evaluation of the volume effects is also needed to understand the inter-tumor variation of the optimal dose. Without a correction for the volume effects in the EUAD, the DNE can be used to analyze the optimal non-uniform dose distribution for the tumor of equal volume. The main goal of this section is to show that the volume of the hypoxic region becomes a parameter that affects the radiobiological effectiveness of dose escalation in the hypoxic regions under certain conditions. Conclusions obtained in this Chapter can be used for dose de-escalation strategies that can reduce dose to normal tissues and normal tissue complications respectively.

The dose escalation to the hypoxic regions is usually governed by a map of the OER derived from PET or MRI images. The LQ cell survival model predicts that the dose escalation in the hypoxic regions should be equal to the level of OER times the dose in oxygenated region to eliminate the radioresistance of hypoxic cells. For example, an OER $= 2$ implies the need for a factor of two in the dose escalation to hypoxic tumor regions to achieve the same level of control as in well-oxygenated tumor regions. The equations derived in this section show that the volume effects in the TCP and non-proliferation of hypoxic cells may alter the need for dose-escalation by a factor equal to the OER to achieve acceptable levels of local tumor control. The TCP dependence on the total and hypoxic tumor volumes derived in Chap. 3 can be used to evaluate the volume effects in the hypoxia-targeted dose escalation.

The TCP is primarily defined by the hypoxic tumor fraction $\mathrm{TCP} \approx \mathrm{TCP}_{\mathrm{hyp}}$ if the hypoxia-targeted dose escalation does not exceed 50% of the dose in oxygenated region and the condition given by Eq. 3.7 is still valid. This case of moderate dose escalation to less than 50% of dose in oxygenated region was typically considered [8]. We evaluate the dependence of hypoxia-targeted dose escalation on the hypoxic tumor volume under assumption of equal TCP. Dose-volume dependence of the TCP described by Eqs. 3.12 and 3.14 can be used for evaluation of dose-volume dependence in the hypoxia-targeted dose escalation Δd:

$$\Delta d = -d_{ox} + OER\frac{-\alpha_{ox} + \sqrt{\alpha_{ox}^2 + 4\beta_{ox}C_{hyp}}}{2\beta_{ox}}, \tag{5.14}$$

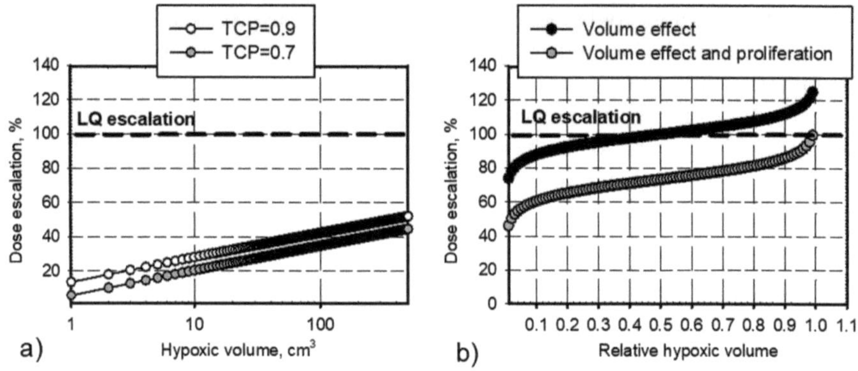

Fig. 5.6 Dependence of hypoxia-targeted dose escalation on hypoxic volume in the head and neck cancer model problem: **a** TCP defined by hypoxic volume (Eq. 5.16); **b** equal TCP in hypoxic and oxygenated sub-volumes computed with and without cell proliferation in the oxygenated sub-volume (Eq. 5.18). A hypothetical dose escalation of 100% computed for OER = 2 in the LQ model is also shown [24]

where d_{ox} is the fractional dose delivered to oxygenated volume and C_{hyp}. is given by [24]:

$$C_{hyp} = \frac{1}{k}\ln\left\{-\frac{\rho V_{hyp,0}}{\ln TCP}\right\}. \tag{5.15}$$

The volume effects in the TCP are computed for the model problem in head and neck cancer described in Chap. 3. The hypoxia-targeted dose escalation computed using Eqs. 5.14 and 5.15 is shown in Fig. 5.6a as a function of the absolute hypoxic volume for different values of the TCP. This dose escalation is smaller than 50% of the dose in oxygenated region for the considered volume range from 1 to 500 °C. The required dose-escalation is also smaller for the smaller absolute hypoxic volumes.

The derived TCP dependence on the total and hypoxic tumor volumes can be used to evaluate the feasibility of different approaches in hypoxia-targeted dose escalation. To eliminate the radioresistance of hypoxic cells, the LQ model predicts that dose escalation in hypoxic regions should be equal to the OER times the dose in oxygenated region. The theoretically required dose escalation is 100% because it was estimated that OER = 2 for conventional fractionation. We have shown that the clinically acceptable values of TCP would require much lower hypoxia-targeted dose escalation (< 50%) with the volume effects considered. The dose-volume effects reduce the dose escalation levels required to overcome radioresistance of hypoxic cells.

5.3.2 Fractional Hypoxic Volume

Equations 5.14 and 5.15 for the dose-volume effects in hypoxia-targeted dose escalation include the density of clonogenic cells that is known with a large uncertainty. There is an alternative derivation of the dose-volume effects based on tumor-average cell surviving fraction. This derivation gives the volume effect dependence on relative hypoxic volume thus excluding the density of clonogenic cells from the calculations [25].

We evaluate the dependence of hypoxia-targeted dose escalation on the relative hypoxic tumor volume under assumption of equal TCP. It follows from Eq. 3.11, that the smaller relative hypoxic volume v_{hyp} would need smaller dose escalation Δd to maintain equal tumor-average cell surviving fraction and the TCP respectively. Using Eq. 1.15 for the cell surviving fraction $S(d_{ox} + \Delta d)$ in Eq. 3.11 for tumor-average cell surviving fraction and applying the standard solution to the final quadratic equation, we can obtain an analytical equation for the dose escalation Δd as a function of relative hypoxic volume. This analytical solution is identical to Eq. 5.14 and where and C_{hyp} should be replaced by the new parameter c_{hyp}:

$$c_{hyp} = \frac{1}{k} \ln\left\{\frac{v_{hyp}}{S_{TA}}\right\}. \tag{5.16}$$

The dose escalation Δd in Eqs. (5.14) and (5.16) can be calculated as a function of fractional hypoxic volume v_{hyp} assuming $S_{\text{TA}} = \text{const}$.

5.3.3 Large Dose Escalation

It is also possible to compute the volume effects when the dose escalation in the hypoxic region exceeds 50% of the dose in well-oxygenated region [24]. In this case, the relationship between the TCP in oxygenated and hypoxic sub-volumes should be considered explicitly. A simple form of equations for volume dependence of dose escalation can be obtained if we assume:

$$TCP_{hyp} = (TCP_{ox})^m, \quad m \geq 1, \tag{5.17}$$

where m is a parameter and TCP_{ox} and TCP_{hyp} are given by Eq. 3.4 and 3.5, respectively. The final equation for dose escalation in this case is equivalent to Eq. 5.12 where the parameter C_{hyp} is replaced by a different parameters c_{hyp}:

$$c_{hyp} = \ln\left[m^{\frac{1}{k}}\left[\frac{V_{hyp,,0}}{V_{tot,0} - V_{hyp,,0}}\right]^{\frac{1}{k}} S_{ox}(d)^{-1}\exp(-\lambda\Delta t)\right]. \tag{5.18}$$

For example, it is useful to consider a special case m = 1 when the radioresistance of the hypoxic volume is eliminated.

The effects of hypoxia-targeted dose escalation larger than 50% could be computed from the TCP relationship in the hypoxic and oxygenated regions given by Eqs. 5.14 and 5.18. In Fig. 5.6b, we show a computed hypoxia-targeted dose escalation as a function of relative hypoxic volume for the equal TCP in oxygenated and hypoxic sub-volumes. One can see that the 100% dose escalation is required only for equal hypoxic and oxygenated volumes if no cell proliferation in the oxygenated volume is considered. A reduction in the relative hypoxic volume below 50% results in reduced dose escalation below 100% and an increase in the relative hypoxic volume above 50% produces an increase in the hypoxia-targeted dose escalation above 100%. One can also see that the inclusion of cell proliferation in the oxygenated region reduces the required hypoxia-targeted dose escalation. This is a result of relative dose de-escalation in the hypoxic region to maintain the TCP_{hyp} equal to the TCP_{ox} in oxygenated region because the hypoxic cells do not proliferate.

Using basic radiobiological equations for the TCP and the cell survival fractions described by the LQ model, we have shown that smaller relative hypoxic volumes require smaller dose escalation to maintain the equal TCP. In general, this effect agrees with the well-known research results that tumors with total smaller volumes require smaller doses to achieve equal TCP because they have a smaller number of clonogenic cells [26]. These results stem from the probabilistic nature of interaction of radiation with matter and agree with the TCP equations described in Chap. 1.

5.3.4 Conclusions

We have shown theoretically that heterogeneous tumors with smaller relative hypoxic volumes may need smaller hypoxia-targeted dose escalation to achieve equal TCP. The smaller doses are needed because the smaller hypoxic volumes have smaller number of clonogenic cells assuming a linear relationship between the number of clonogenic cells and the hypoxic volume. This volumetric effect is observed at the dose escalation levels when the hypoxic region remains radioresistant compared to oxygenated region and primarily defines TCP. It is explained by dependence of average tumor cell survival only on the hypoxic volume for a moderate dose escalation.

Theoretically, to eliminate radioresisance of hypoxic cells, the dose escalation in hypoxic regions should be done to the level of OER times the dose in oxygenated region. It was estimated that OER = 2 for conventional fractionation and OER = 3 for the hypofractionated dose regimens [27]. However, maintaining the normal tissue complications at an acceptable level is another challenging issue in using hypoxia-targeted dose escalation; therefore, the normal tissue constraints limit the dose escalation levels to about 50% of the standard prescription dose [4]. Dose escalation levels of about 50% or less, which were assumed in most dose escalation studies satisfy these conditions. This moderate dose escalation increases cell killing in the hypoxic region; however, the hypoxic region remains more radioresistant than the oxygenated region and defines the TCP.

The results of this section agree with previously published research where it was shown that an increase in initial tumor volume leads to decreased TCP for the equal dose for some tumors [26]. If a tumor has a radioresistant hypoxic region with relative volume larger than a few percent, this hypoxic region primarily defines TCP. Therefore, it can be expected that larger hypoxic volumes would require larger hypoxia-targeted dose escalation to achieve equal TCP.

The results of this research can be used for dose de-escalation strategies in dose painting techniques that may reduce dose to normal tissue and normal tissue complications respectively without deterioration of tumor control. The practical application of the proposed approach for the evaluation of dose escalation may require some additional modifications. For example, our calculations were based on the constant value of OER and dose in the hypoxic region. If OER and dose vary in the hypoxic region, average cell survival should be computed using the integration method.

Our analysis indicates that the dose to oxygenated cells could be reduced without adversely affecting the TCP. Clinical data also support the hypothesis that dose de-escalation could be feasible if the hypoxic region is completely re-oxygenated before the end of treatment [9]. Robust practical implementation of this approach is more challenging than the dose escalation to hypoxic regions because the TCP can be compromised by possible delivery errors where hypoxic cells could be irradiated with the de-escalated dose. However, an effective approach was proposed where total dose was de-escalated uniformly for the entire tumor in patients with resolution of hypoxia on repeat mid-treatment FMISO PET scans [15]. This approach produced substantially reduced normal dose toxicity without deterioration of tumor control.

References

1. Rajendran JG, Schwartz DL, O'Sullivan J, Peterson LM, Ng P, Scharnhorst J, Grierson JR, Krohn KA. Tumor hypoxia imaging with [F-18] fluoromisonidazole positron emission tomography in head and neck cancer. Clin Cancer Res. 2006; 12(18): 5435–41.
2. Spence AM, Muzi M, Swanson KR, O'Sullivan F, Rockhill JK, Rajendran JG, Adamsen TC, Link JM, Swanson PE, Yagle KJ, Rostomily RC, Silbergeld DL, Krohn KA. Regional hypoxia in glioblastoma multiforme quantified with [18F] fluoromisonidazole positron emission tomography before radiotherapy: correlation with time to progression and survival. Clin Cancer Res. 2008; 14:2623–30.
3. Mayr NA, Huang Z, Wang JZ, Lo SS, Fan JM, Grecula JC, Sammet S, Sammet CL, Jia G, Zhang J, Knopp MV, Yuh WT. Characterizing tumor heterogeneity with functional imaging and quantifying high-risk tumor volume for earlier prediction of treatment outcome: cervical cancer as a model. Int J Radiat Oncol Biol Phys. 2012; 83(3):972–979.
4. Sovik A, Malinen E, Bruland OS, Bentzen SM, Olsen DR. Optimization of tumor control probability in hypoxic tumors by radiation dose redistribution. Phys Med Biol. 2007; 52: 499–513.
5. Axente M, Lin P-S, Pugachev A. Single-fraction simulation of relative cell survival in response to uniform versus hypoxia-targeted dose escalation. Phys Med Biol. 2012; 57: 2757–2774.

6. Chvetsov AV, Rajendran JG, Zeng J, Patel SA, Bowen SR, Kim EY. Theoretical effectiveness of cell survival in fractionated radiotherapy with hypoxia-targeted dose escalation. Med Phys. 2017; 44(5):1975–1982.
7. Chvetsov AV, Pugachev A. Biological effectiveness of uniform and nonuniform dose distributions in radiotherapy for tumors with intermediate oxygen levels. Biomed Phys Eng Express, 2024; 10: 065048.
8. Sövik A, Malinen E, Olsen D. Strategies for biologic image guided dose escalation: a review. Int J Radiat Oncol Biol Phys. 2009, 73: 650–658.
9. Das SK, Ten Haken RK, Functional and molecular image guidance in radiotherapy treatment planning optimization, Semin Radiat Oncol. 2011, 21(2): 111–118.
10. Bentzen SM, Gregoire V. Molecular imaging-based dose painting: a novel paradigm for radiation therapy prescription. Semin Radiat Oncol. 2011; 21(2):101–10.
11. Salem A. Hypoxia-Targeted Dose Painting in Radiotherapy. Semin Radiat Oncol. 2023, 33: 298–306.
12. Welz S, Paulsen F, Pfannenberg C, Reimold M, Reischl G, Nikolaou K, La Fougère C, Alber M, Belka C, Zips D, Thorwarth D. Dose escalation to hypoxic subvolumes in head and neck cancer: A randomized phase II study using dynamic [^{18}F]FMISO PET/CT. Radiother. Oncol. 2022; 17: 30–36.
13. Schiavo F, Toma-Dasu I, Lindblom EK. Hypoxia dose painting in SBRT - the virtual clinical trial approach. Acta Oncol. 2023; 62: 1239–124.
14. Chvetsov AV, Muzi M. Equivalent uniform aerobic dose in radiotherapy for hypoxic tumors. Phys Med. Biol. 2024; 69: 085011.
15. Lee N, Schoder H, Beattie B, Lanning R, Riaz N, McBride S, Katabi N, Li D, Yarusi B, Chan S, Mitrani L, Zhang Z, Pfister DG, Sherman E, Baxi S, Boyle J, Morris LG, Ganly I, Wong R, Humm J. Strategy of Using Intratreatment Hypoxia Imaging to Selectively and Safely Guide Radiation Dose De-escalation Concurrent with Chemotherapy for Locoregionally Advanced Human Papillomaviruse - Related Oropharyngeal Carcinoma, Int J Radiation Oncol Biol Phys. 2016; 96(1): 9–17.
16. Ruggieri R, Naccarato S, Nahum AE. Severe hypofractionation: Non-homogeneous tumor dose delivery can counteract tumor hypoxia. Acta Oncologica. 2010; 49: 1304–1314.
17. Ruggieri R, Nahum AE. The impact of hypofractionation on simultaneous dose-boosting to hypoxic tumor subvolumes. Med Phys. 2006; 33(11): 4044–4055.
18. Webb S, Evans PM, Swidell W, Deasy JD. A proof that uniform dose gives the greatest TCP for fixed integral dose in the planning target volume. Phys Med Biol. 1994; 39: 2091–2098.
19. Niemierko A. Reporting and analyzing dose distributions: A concept of equivalent uniform dose. Med Phys. 1997; 24: 103–110.
20. Wigg DR. Applied Radiobiology and Bioeffect Planning. Madison, WI. Medical Physics Publishing; 2001.
21. Koh WJ, Bergman KS, Rasey JS, Peterson LM, Evans ML, Graham MM, Grierson JR, Lindsley KL, Lewellen TK, Krohn KA, Griffin TW. Evaluation of oxygenation status during fractionated radiotherapy in human non-small cell lung cancer using (F-18) fluromisonidazole positron emission tomography. Int J Radiat Oncol Biol Phys. 1995; 33(2): 391–398.
22. Chvetsov AV, Sandison GA, Schwartz JL, Rengan R. Ill-posed problem and regularization in reconstruction of radiobiological parameters from serial tumor imaging data. Phys Med Biol. 2015; 60(21): 8491–8503.
23. Chvetsov AV, Palta JR, Nagata Y. Time-dependent cell disintegration kinetics in lung tumors after irradiation. Phys Med Biol. 2008; 53: 2413–2423.
24. Chvetsov AV, Stewart RD, Kim M, Meyer J, Rengan R. Volume effects in the TCP for hypoxic and oxygenated tumors. Med Phys. 2020; 47: 4626–4633.

25. Chvetsov AC, Zeng J, Rajendran JR. Volume dependence in hypoxia-targeted dose escalation. Med Phys. 2018; 45(11): 5325–5331
26. Brenner DJ. Dose, volume, and tumor-control predictions in radiotherapy. Int J Radiat Oncol Biol Phys. 1993; 26: 171–179.
27. Hall EJ, Giaccia AJ. Radiobiology for the Radiologist. 6th ed. Philadelphia, PA: Lippincott Williams & Wilkins; 2006.

Chapter 6
Stability of Inverse Planning with Different Objective Funcrions

6.1 Dose Optimization for Hypoxic Tumors

6.1.1 Mechanistic and Empirical Biological Objective Functions

Planning the biological effects using the radiotherapy optimization based on objective functions derived from the TCP has been suggested as a mechanism for individualizing therapy and improving outcomes [1–4]. Treatment planning studies usually show improvements of the TCP; however, the TCP application in biological optimization poses a challenging task due to the large number of radiobiological parameters and their uncertainties. For example, computing the TCP would require the 3D distribution of the density of tumor clonogenic cells that is usually not known. The hypoxia targeted dose escalation based on the TCP maximization was investigated in theoretical treatment planning studies by was mostly not verified using treatment outcomes analysis. Therefore, the TCP based optimization did not find application in the clinic and the dose escalation studies based on the functional images mostly include empirical approaches.

The concept of EUAD that is an intermediate concept between the physical dose and the TCP, can be considered as a more robust and practical approach for optimization of biological effects. In Chap. 5, we have shown that the radiotherapy dose redistribution with higher doses delivered to regions with higher OER and lower doses delivered to regions with lower OER may produce higher values of the EUAD. However, we have used approximate empirically derived dose distributions. In this Chapter, we describe accurate dose optimization methods to compute the optimal OER-based dose distributions using direct optimization of the EUAD concepts. The EUAD concept is needed to derive non-uniform dose distributions in the hypoxic tumors that are expected to improve the Therapeutic Ratio of radiation therapy.

A. V. Chvetsov, *Optimization Problems in Radiotherapy for Hypoxic Tumors*, Biological and Medical Physics, Biomedical Engineering,
https://doi.org/10.1007/978-981-96-7001-7_6

Application of hypoxia imaging in accurate biological dose optimization requires application of quantitative prognostic factors such as TCP or EUAD that can predict the non-linear radiotherapy response. However, the hypoxia parameters derived from hypoxia imaging could be applied only in empirical biological dose optimization algorithms. It is possible to perform empirical biological optimization where the radiotherapy doses are empirically selected as a function of volumes and uptake of the hypoxia biomarkers [1], These imaging derived hypoxia parameters do not include the radiotherapy dose regimens, and the non-linearity of radiation response described by the LQ cell survival model. Therefore, the imaging-based hypoxia parameters may have limited prognostic value and cannot be used effectively in dose optimization where a relationship between treatment outcomes and three-dimensional dose distributions is needed. However, the mathematical dose optimization algorithms described in this Chapter, can be applied to both mechanistical and empirical objective functions.

6.1.2 Mathematical Problems in Dose Optimization

To proceed with dose optimization for hypoxic tumors based on the EUAD, we need to define the basic equations usually used in inverse treatment planning for radiation therapy. The two major components of inverse planning for radiation therapy are the dose calculation algorithm and the optimization algorithm for solving the *radiation transport inverse problem*. The solution of the optimization problem is typically done by constrained iterative techniques to find the global extreme of an objective function. The optimized beam intensity functions that produce the optimal dose distributions are obtained using algorithms of inverse treatment planning [5, 6].

Objective functions in treatment planning relate to absorbed dose or radiobiological response. Constraints, both physical and practical, may be expressed as equalities or inequalities that effectively restrict the space of solutions for the optimization algorithm. An example of a physical constraint is the fact that beam intensities cannot be negative. Practical constraints might arise from the beam delivery technology available for treatment such as multi-leaf collimation or the available treatment time allocated per patient to implement many port plans.

The inverse treatment planning is an inverse radiation transport problem known to be poorly conditioned in the sense it may become unstable due to small perturbations in the input data. Inverse source radiation transport problem is governed by a Fredholm integral equation of the first kind that is notoriously ill-conditioned [7, 8]. Mathematical optimization of intensity-modulated involves large number free variables. Usually, these variables are non-uniform intensities. If, in addition, we also include the beam energy and beam orientation the number of variables will increase significantly. The only practical way to handle such many variables is by efficient numerical optimization methods. The numerical solution may become unstable simply from the truncation and approximation errors in the numerical algorithm. The ill-conditioning of the inverse treatment planning can lead to solutions with non-physical oscillation in beam intensities. This, in turn, will slow down the convergence rate of iterative methods used to solve optimization process and create problems with the beam delivery procedure.

To stabilize the numerical solutions, special regularization techniques have been developed [7, 8]. We customize these techniques for inverse treatment planning algorithms to improve quality of numerical solutions and, in turn, the modality of dose delivery of to the cancer tumor. Research in inverse treatment planning usually addresses two key topics:

(1) selection of objective function that produces solutions with desirable properties.
(2) development of efficient iterative algorithms to find these solutions quickly.

In this Chapter we address the first topic.

6.2 Inverse Treatment Planning

6.2.1 Ill-Posed Inverse Radiation Transport Problem

Inverse treatment planning is a kind of inverse source radiation transport problem which is based on numerical solution for the Fredholm integral equation of the first kind [9]:

$$D(\boldsymbol{r}) = \int_a^b f(\xi) d(\boldsymbol{r}, \xi) d\xi, \tag{6.1}$$

where $D(\boldsymbol{r})$ is the desirable distribution of the absorbed dose in tumors, $f(\xi)$ is the beam intensity function and $d(\boldsymbol{r},\xi)$ is the dose distribution from a narrow beam (“beamlet”). The methods of inverse planning utilize a discretized form of Eq. 6.1 given by

$$D(\boldsymbol{r}) = \sum_{m=1}^{M} w_m d_m(\boldsymbol{r}), \tag{6.2}$$

that can be obtained for the case of degenerated kernel:

$$d(\boldsymbol{r}, \boldsymbol{\xi}) = \sum_{m=1}^{M} d_m(\boldsymbol{r}) Y_m(\boldsymbol{\xi}). \tag{6.3}$$

The weight w_m in Eq. 6.2 is an integral of the product of the beam intensity function $f(\xi)$ and the weighting function $Y_m(\xi)$.

Equation 6.1 is unstable on the dose distribution $D(\boldsymbol{r})$ and, therefore, is ill-posed. To show this, we introduce a high frequency perturbation:

$$\delta f(\boldsymbol{\xi}) = \exp(i\omega\xi), \quad \omega \geq 0 \tag{6.4}$$

with a limited amplitude into the beam intensity $f(\xi)$. Substituting this perturbation into Eq. 6.1 and integrating by parts, we obtain that the corresponding dose perturbation:

$$\delta D(\boldsymbol{r}) = \int_a^b \exp(i\omega\xi) d(\boldsymbol{r}, \xi) d\xi = \frac{1}{i\omega} \exp(i\omega\xi) d(\boldsymbol{r}, \xi) \Big|_a^b - \frac{1}{i\omega} \int_a^b \frac{\partial d(\boldsymbol{r}, \xi)}{\partial \xi} \exp(iw\xi) d\xi = O\left(\frac{1}{\omega}\right) \tag{6.5}$$

that can be infinitesimal for high frequency ω. Therefore, infinitesimal perturbations in dose $D(\boldsymbol{r})$ may produce large perturbations in beam intensity $f(\xi)$ that makes the problem ill-posed. Even if the dose function $D(\boldsymbol{r})$ is defined accurately as in radiotherapy optimization problems, the approximation errors in the numerical algorithm can produce large non-physical oscillations in the beam intensity $f(\xi)$.

Before applying a regularization technique to correct for the property of ill-posedness in Eq. 6.2, it is useful to analyze the features of the desirable dose distribution $D(\boldsymbol{r})$ used in inverse planning and how it may influence the optimization process. Unlike some other inverse radiation transport problems, for instance, in nuclear medicine or spectrometry, where the objective function is based on physical measured data, the desirable dose distribution $D(\boldsymbol{r})$ in inverse treatment planning is an artificially created mathematical concept. We would like to underline the following features of $D(\boldsymbol{r})$ which may affect the instabilities and methods for their correction: (1) Firstly, as we already mentioned, the desirable dose distribution does not have fluctuations, usually referred as "noise", associated with the uncertainties of measurement. This practically removes one of the most important sources of fluctuations in the reconstructed function thus making inverse planning less prone to instabilities. (2) Secondly, the desirable dose distribution $D(\boldsymbol{r})$ usually has breaks at the boundaries between PTV, critical structures and normal tissue. For instance, the function $D(\boldsymbol{r})$ has a jump:

$$\Delta D = |D_{PTV} - D_{Normal\ Tissue}| \tag{6.6}$$

at every boundary point between PTV and normal tissue that is due to the dose prescription. These non-physical jumps in dose distribution $D(\boldsymbol{r})$ cannot be reproduced by a linear combination of beamlet dose functions that are smooth physical concepts based on the scattering cross-sections. Therefore, if we substitute a desirable dose distribution $D(\boldsymbol{r})$ that is an artificial mathematical concept and a beamlet dose function $d_m(\boldsymbol{r})$ that is a physical concept into the discrete Eq. 6.2, an exact solution to this equation cannot be found because of the physical limitations. If inverse planning is considered as an optimization problem this results in a non-removable discrepancy between the exact mathematical concept and physically achievable solution for the desirable dose distribution $D(\boldsymbol{r})$. Practically, this means that the objective function will be a certain value that never approaches zero. This feature is important for regularization techniques that are based on the perturbation of the objective function by a small regularization term.

6.2.2 *Regularization Algorithms*

The issues of stability and accuracy of inverse treatment planning for intensity modulated radiation therapy (IMRT) are mostly related to inverse planning belonging to the class of *discrete ill-posed problems* with underlying Fredholm integral equation of the first kind. This results in non-physical oscillations in optimized beam intensity functions. The instabilities can be controlled in practice by smoothing the optimized beam intensity patterns by filtering when the optimization is completed (this procedure can be also applied iteratively during optimization) [10, 11].

However, a more accurate solution to this problem is a regularization technique where the ill-posed problem is modified to obtain a well-posed problem. The application of regularization techniques improves the dose uniformity in the planning target volumes and decreases delivery time which are the desirable characteristics of IMRT [12]. Several regularization methods have been developed to improve the quality of numerical solutions of Eq. 6.1: (1) regularization by filtering, (2) iterative regularization methods, and (3) variational regularization methods [13, 14].

(a) **Regularization by Filtering**

Probably, the first regularization techniques applied to inverse treatment planning have been regularization by filtering in the direct Fourier transformation [15] and iterative regularization methods. Regularization by filtering did not find broad application in clinical inverse treatment planning and was restricted primarily to research applications. This may be attributed to the fact that the regularization by filtering is not practical for very large ill-conditioned systems encountered in radiation therapy optimization problems.

(b) **Iterative Regularization Methods**

The iterative regularization methods are based on the minimization of a functional that relates to the least squares fit-to-data functional, and they can be effectively applied for very large ill-conditioned systems. The regularization in these methods is achieved using a special iteration scheme in minimum search. There are several iterative methods available, for instance, the classical iterative schemes of Cimmino and Landweber or conjugate gradient method. The Cimmino algorithm was, probably, the most researched approach for practical applications in optimization problems of inverse treatment planning [15–19]. The classical iterative methods, as Cimmino or Landweber algorithms, are usually characterized by slow convergence of the iteration process, therefore, they require many iterations to generate accurate smooth solutions. Several acceleration methods are available [13] and other acceleration methods have been proposed [19]. We note that the computational cost of Cimmino algorithm in inverse treatment planning problems is quite acceptable despite the large number of iterations (500 or more against 25–50 iterations for the sequential quadratic programming or quasi-Newton algorithms) [18]. This may be attributed to the low cost of one iteration and effective computer implementation.

(c) **Variational Regularization Methods**

The variational regularization technique has also been applied to inverse treatment planning for IMRT [9, 20, 21]. The regularization effect in this method is achieved by modification of the least squares objective function with a stabilizing functional and this approach has many advantages that are discussed below. A few approaches to this method have been proposed, for example, one of these approaches utilized zeroth order regularization with target values evaluated using a polynomial filter [21] and another approach is based on an analogy with membrane of a minimum surface [20]. These methods are based on many problem-dependent computational parameters; therefore, these approaches are efficient for some problems; for others, they may not. The latter approach introduces a regularizing term that includes the second power of the second derivative and, in this sense, is close to the second order regularization. Using derivatives has some advantages compared to using filters. Firstly, a derivative is a universal characteristic of a function and, secondly, it can be evaluated locally using small number of adjacent points (only two points for the first order derivative). However, using the second derivative may result in excessive smoothing of the solution. Another deficiency of this formulation is that it may not provide the uniform convergence of solutions because the function and its first order derivative are missing in the numerator of the regularization term.

In contrast to these problem-specific approaches, there is a more universal form of objective function for the variational regularization method that regularizes the numerical solutions and maintains the best possible approximation of the exact solution. This approach takes advantage of theoretical development of accurate methods for ill-posed problems and these objective functions have been independently developed by Tikhonov and Phillips, but usually referred in the literature as Tikhonov regularization [7, 8]. This approach is well-posed and allows one to control convergence of all derivatives up to the n-th order. It should be noted that this approach offers a universal computational tool for the inverse problems. Previously, this regularization method was applied for the reconstruction of electron spectra from depth-dose distributions that also belong to the class of inverse radiation transport problems and was found it is very efficient in both suppressing non-physical oscillations and maintaining a good approximation. In Chap. 2, we successfully used Tikhonov's regularization for solving the inverse problem related to the reconstruction of radiobiological parameters from serial tumor imaging data.

6.2.3 Variational Regularization

There are many practical advantages associated with variational regularization. Firstly, the variational regularization method is universal with respect to the minimum search: any deterministic or statistical optimization method can be used to find a minimum of the modified objective function. Some of these optimization methods are extremely fast as quasi-Newton methods. Others optimization methods such as simulated annealing may have the ability to find the global minimum. We can always choose the optimization method suitable for solving a particular problem. Second, the objective function in variational methods can be easily modified to obtain a solution with desirable properties, for instance, to include biological or functional information. Third, the variational regularization method can be easily implemented into clinical and research optimization codes which have been already developed.

In the variational regularization method, a quadratic objective function is modified to achieve stability of numerical solutions by adding a stabilizing term. The modified objective function can be written as

$$\theta_n[f(\xi)] = \int I(\boldsymbol{r})(D(\boldsymbol{r}) - D_p(\boldsymbol{r}))^2 d\boldsymbol{r} + \alpha\Omega_n[f(\xi)], \tag{6.7}$$

where $D_p(\boldsymbol{r})$ is the prescribed dose distribution, $I(\boldsymbol{r})$ is the importance factor, $\Omega_n[f(\xi)]$ is the stabilizing functional and α is the regularization parameter. The integration in Eq. 6.7 can be performed over a two- or three-dimensional region in our notation. However, it should be considered that the intensity function $f(\xi)$ is a two-dimensional function for the three-dimensional case. A non-negative intensity function $f_\alpha(\xi)$ minimizing the objective function $\theta_n[f(\xi)]$ will give a solution to the inverse treatment planning problem:

$$f_\alpha(\xi) = \arg\min\{\theta_n[f(\xi)]\}. \tag{6.8}$$

If the regularization parameter α is small, a solution obtained using the objective function given by Eq. 6.8 can be considered as an approximation to the solution to Eq. 6.1.

The Tikhonov's regularization functional $\Omega_n[f(\xi)]$ in Eq. 6.7 is given by [7, 8]

$$\Omega_n[f(\xi)] = \int_a^b d\xi \sum_{n=0}^{N} p_n(\xi)\left[\frac{d^{(n)}f(\xi)}{d\xi^{(n)}}\right]^2 \tag{6.9}$$

where $p_n(\xi) > 0$ is a positive continuous weighting function. The stabilizing functional with $n = 0$ provides the weak regularization with only mean squared convergence $f_\alpha(\xi) \longrightarrow f(\xi)$, the stabilizing functional with $n = 1$ provides the strong regularization with uniform convergence $f_\alpha(\xi) \longrightarrow f(\xi)$, and the stabilizing functional with $n > 1$ provides the n-order regularization with uniform convergence of function and its derivatives up to the order $n - 1$. The n-th derivative exhibits mean squared convergence for in the latter case. Mean squared convergence assumes convergence of the norm:

$$\|f_\alpha(\xi)\|_{L_2} = \sqrt{\int |f_\alpha(\xi)|^2 d\xi} \tag{6.10}$$

Uniform convergence assumes that for any $\varepsilon > 0$ we can find $\alpha_0(\varepsilon)$ such that

$$\|f_\alpha(\xi) - f(\xi)\| \leq \frac{\varepsilon}{2} \quad for\, \alpha \leq \alpha_0(\varepsilon) \tag{6.11}$$

The regularization follows from the fact that the optimization problem given by Eqs. (6.7)–(6.9) with $\alpha \neq 0$ is equivalent to the Fredholm integral equation of the second kind which is well-posed. We present the main results of this derivation. Let us consider a variation of the objective function $\theta_n[f(\xi)]$ given by Eq. 6.7 with respect to the beam intensity function $f(\xi)$. At the minimum of the objective function $\theta_n[f(\xi)]$ its variation is equal zero $\delta\theta_n[f(\xi)] = 0$ thus we obtain the Euler equation

$$\alpha \sum_{n=0}^{N} \int_a^b p_n(\xi) f^{(n)}(\xi) \delta f^{(n)}(\xi) d\xi$$
$$+ \int d\boldsymbol{r} I(\boldsymbol{r}) \left\{ \int_a^b d(\boldsymbol{r},\eta) f(\eta) d\eta - D_p(\boldsymbol{r}) \right\} \int_a^b d(\boldsymbol{r},\xi) f(\xi) d\xi = 0. \tag{6.12}$$

Performing integration by parts in the first term and further substituting the variation $\delta f(\xi)$ as a delta-function, we obtain an integro-differential equation:

$$\alpha \sum_{n=0}^{N} (-1)^n \frac{d^n}{d\xi^n} \left[p_n(\xi) f^{(n)}(\xi) \right] + \int_a^b K(\boldsymbol{\xi},\eta) f(\eta) d\eta = R(\xi), \tag{6.13}$$

where the kernel $K(\xi,\eta)$ and the function $R(\xi)$ are given by

$$K(\xi,\eta) = \int d\boldsymbol{r} I(\boldsymbol{r}) d(\boldsymbol{r},\eta) d(\boldsymbol{r},\xi), \tag{6.14}$$

$$R(\xi) = \int d\boldsymbol{r} I(\boldsymbol{r}) D_p(\boldsymbol{r}) d(\boldsymbol{r},\xi). \tag{6.15}$$

Formulation of the optimization problem in the form of Eq. 6.13 presents proof that the regularized optimization problem is well-posed. Let us suppose for mathematical brevity $p_n(\xi) = 1$ and consider the case $n = 0$. Then, Eq. 6.13 becomes the Fredholm integral equation of the *second kind*

$$\alpha f(\xi) + \int_a^b K(\boldsymbol{\xi},\eta) f(\eta) d\eta = R(\xi). \tag{6.16}$$

For the case of $n > 0$, Eq. 6.13 is expressed as the Fredholm integral equation of the second kind using formalism of Green function. The intensity function $f(\xi)$ outside the integration sign is the major difference between the Fredholm integral equations of the first kind given by Eq. 6.1 and the second kind given by Eq. 6.16. The intensity function outside the integration sign removes the discontinuous dependence of $f(\xi)$ on the function $R(\xi)$ thus making the problem well-posed.

There are two key problems in the variational regularization methods: (1) selection of the stabilizing functional and (2) evaluation of an optimal value of the regularization parameter—a weighting coefficient for the stabilizing functional. The research on inverse planning addressed primarily the selection of the stabilizing functional [9, 20, 21]. The regularization parameter was considered as a fitting coefficient that was selected, for instance, using visual control to separate noise from signal [9]. In this Chapter, we apply a method of L-curve to evaluate the optimal value of regularization parameter in the radiotherapy optimization problems [13, 14].

6.2.4 *Stabilizing Functionals*

Using the norm defined by Eq. 6.10 and assuming $p_n(\xi) = 1$, the stabilizing functions with the zeroth (n = 0), first (n = 1) and second (n = 2) order regularization can be written in compact form as:

$$\Omega_0[f(\xi)] = \|f(\xi)\|_2^2, \tag{6.7}$$

$$\Omega_1[f(\xi)] = \|f(\xi)\|_2^2 + \left\|\frac{df(\xi)}{d\xi}\right\|_2^2, \tag{6.18}$$

$$\Omega_2[f(\xi)] = \|f(\xi)\|_2^2 + \left\|\frac{df(\xi)}{d\xi}\right\|_2^2 + \left\|\frac{d^2f(\xi)}{d\xi^2}\right\|_2^2. \tag{6.19}$$

The regularization properties of these functions are studied in [9]. As we already mentioned, the stabilizing functional with $n = 0$ provides weak regularization with mean squared convergence $f_\alpha(\xi) \longrightarrow f(\xi)$, Therefore, the zeroth order regularization cannot suppress high-frequency oscillations even for extremely high value of regularization parameter.

The stabilizing functional with n = 1 provides strong regularization with uniform convergence $f_\alpha(\xi) \longrightarrow f(\xi)$, and the stabilizing functional with n > 1 provides the n-order regularization with uniform convergence of function and its derivatives up to the order n − 1. Therefore, these high-frequency oscillations are suppressed by the first and second order regularization. To choose between the first and the second order regularization, we must compare the beam profiles with the smallest value of α where the high-frequency fluctuations are completely removed. Comparing the profiles shows that the second order regularization may lead to the excessive smoothing of intensity profile. Therefore, the first order regularization (n = 1) would be the most optimal case for radiotherapy dose optimization.

6.2.5 Algorithms for Minimum Search

(a) **Gradient Based Optimization Algorithms**

There are several efficient methods have been developed for finding the minimum of the objective function given by Eq. 6.7 depending on the requirements of optimizatation. For example, the minimum of the objective function given by Eq. 6.7 can be found using the L-BFGS-B code that has been developed at the Optimization Technology Center, a joint venture of Argonne National Laboratory and Northwestern University [22]. The code is based on the L-BFGS-B algorithm which is a limited-memory version of the quasi-Newton method with simple constraints on the variables. The method is convenient for practical use because it requires only an objective function and its derivatives as input to perform the minimization search. A comparison of various optimization schemes has shown that this algorithm is the most efficient in terms of computing time, especially for large problems encountered in radiation therapy. The main advantages of the L-BFGS-B algorithm are its low computational cost per iteration, its modest memory requirements, and its ability to ability to solve problems in which the Hessian matrices are large, unstructured, dense, and unavailable.

(b) **Simulated Annealing**

A gradient-based algorithm requires estimation of gradients for a complex objective function and could be trapped in the local minima. These disadvantages can be improved using the simulated annealing that is a global optimization method and can distinguish between different local optima. Since it can find the global minimum of a complex objective function without evaluating its gradients, the simulated annealing algorithm would be the best choice for radiotherapy optimization problems. Since the algorithm makes very few assumptions about the function to be optimized, it is quite robust with respect to non-quadratic surfaces. In fact, simulated annealing can be used as a local optimizer for difficult functions. The optimization algorithms based on the simulated annealing are usually associated with longer computational times compared to the gradient based methods. As an example of simulation annealing algorithms, we mention an implementation of simulated annealing [23] that was based on the continuous simulated annealing global optimization algorithm described in [24].

The variational regularization technique can be applied in connection with any statistical or deterministic method to minimize the quadratic objective function. It should be emphasized that the instabilities relate solely to the Fredholm integral equation of the first kind which governs the optimization search, therefore, the effect of using the regularization technique would be similar when using other gradient based or statistical optimization methods.

6.3 L-Curve Criterion for Regularization Parameter

6.3.1 L-Curve Analysis

Selection of optimal regularization parameter α is one of the most difficult components of the variational regularization approach. Smoothing effect is not our only goal. We would like the smooth solution to be as close as possible to the exact solution. A regularization parameter is optimal if the smoothest solution that satisfies the least squares objective function is taken as an optimal solution. To select an optimal value of regularization parameter in the optimization problems for IMRT with variational regularization, we apply the L-curve method that was developed to determine the regularization parameter in the discrete ill-posed problems [13, 25]. It is also useful to mention other possible applications of the variational regularization method and the L-curve analysis in the radiotherapy physics that may include reconstruction of radiobiological parameters from serial tumor imaging (considered in Chap. 2) and reconstruction of electron spectra from measured PDDs [26] that are also inverse problems like the inverse treatment planning. Another application of the L-curve method to the choice of the regularization parameter for perfusion quantification with MRI [27].

The L-curve method is based on finding the regularization parameter that minimizes the residual norm that is a measure of accuracy of fit and the solution norm that is a measure of smoothness of solution. The L-curve method was developed for determination of the optimal value of regularization parameter in discrete ill-posed problems; therefore, it can be applied to inverse treatment planning [28]. For further consideration we formulate the optimization problem for inverse treatment planning given by Eqs. 6.7–6.9 in the following compact form:

$$w_\alpha = \arg\min\left\{ \left\| \sqrt{\boldsymbol{I}}\big(D(\boldsymbol{r}) - D_p(\boldsymbol{r})\big) \right\|_2^2 + \alpha\Omega_n \right\}, \tag{6.20}$$

and introduce the following variables:

$$\xi(\alpha) = \left\| \sqrt{\boldsymbol{I}}\big(D(\boldsymbol{r}) - D_p(\boldsymbol{r})\big) \right\|_2, \tag{6.21}$$

$$\eta(\alpha) = \sqrt{\Omega_n}. \tag{6.22}$$

The first variable $\xi(\alpha)$ describes the accuracy of fit of the calculated dose to the prescribed dose distribution. The second variable $\eta(\alpha)$ relates to the stabilizing functional Ω and presents a norm of smoothness of solution. The optimal value of the regularization parameter α is the one which minimizes both variables $\xi(\alpha)$ and $\eta(\alpha)$.

The main idea of the L-method is to plot the smoothing norm $\eta(\alpha)$ as a function of the residual norm $\xi(\alpha)$ for all values of the regularization parameter α. This characteristic curve has an L-shaped dependence, and the optimal value of regularization parameter can be found at the "corner" of the L-curve. Applying formalism of the singular value decomposition for the Tikhonov regularization, it can be shown that the smoothing norm $\eta(\alpha)$ is a monotonically decreasing continuous function of the residual norm $\xi(\alpha)$ [13, 14, 25]. Remarkably, the function has an L-shaped form with distinctive vertical and horizontal parts. Of practical importance is the vertical part of the L-curve because the smoothing norm $\eta(\alpha)$ is decreasing and the residual norm $\xi(\alpha)$ remains approximately of the same value. It follows from the vertical part that we can reduce the smoothing norm practically without changing the residual norm. However, after passing the "corner" of the L-curve the situation will be different – small changes in the smoothing norm will cause large changes in the residual norm.

The distinctive L-shape of the L-curve is a fundamental property that is defined by the ill-posedness of the Fredholm integral equation of the first kind [13]. The goal of the L-curve analysis is to compute a simultaneous minimum of the smoothing and residual norms that is found at the corner of the L-curve. The L-curve analysis for the discrete ill-posed problems involves optimization of two criteria. The first criterion is the smoothing norm that is the measure of smoothness of solution and second criterion is the residual norm that is the measure of accuracy of fit to the optimization goals. Therefore, the L-curve analysis can be formally viewed as a multi-criteria optimization which is currently studied for IMRT applications [29].

6.3.2 *Maximum Curvature of L-Curve*

The optimal value of the regularization parameter can be found in the vicinity of the "corner" of the L-curve where both values of smoothing norm $\eta(\alpha)$ and residual norm $\xi(\alpha)$ achieve their minimum. For an idealized L-curve, this point would be exactly at the corner. For the most real cases, the corner of the L-curve is a region. Thus, we must introduce a criterion to select a point from this region. For instance, the corner of the L-curve has the maximum curvature, and the regularization parameter can be found at the point of maximum curvature. The criterion of maximum curvature is the most researched and we apply this criterion in this book [13, 14, 25]. To apply this criterion, we must calculate the curvature $k(\alpha)$ of the L-curve which is given by

$$k(\alpha) = \frac{\xi'_\alpha \eta''_\alpha - \xi''_\alpha \eta'_\alpha}{\left((\xi'_\alpha)^2 + (\eta'_\alpha)^2\right)^{\frac{3}{2}}}, \tag{6.23}$$

where differentiation is with respect to α. Then, the regularization parameter is selected at the point where the curvature function $k(\alpha)$ achieves its maximum.

Although the L-curve has a simple shape and the point of maximum curvature can be determined intuitively, computing this point in a robust way is not an easy task. The regularization parameter approaches zero $\alpha \to 0$ and the first derivative approaches infinity $\eta'_{\xi} \to 0$ on the vertical part of the L-curve which complicates the numerical analysis. We discuss several methods for computing the point of maximum curvature for inverse treatment planning problems [28].

To compute the point of maximum curvature, we must know at least a few points on the L-curve thus the optimization problem must be solved several times. This approach cannot be applied effectively for clinical inverse treatment planning where the calculational time is critical. Another way is to use the approximate methods for evaluation of the L-curve, for instance, calculate an L-ribbon that includes the L-curve [30]. However, these methods will also increase the computational time and introduce inevitable error in determining the regularization parameter.

Therefore, we believe that prior knowledge is the best way to utilize advantages of the L-curve method for inverse treatment planning. The L-curve and the regularization parameter should be determined during commissioning of a treatment planning system using representative clinical cases, for instance, for the head-and-neck or prostate cancer treatment. The validity of this approach can be explained by the following facts. Usually, the value of the regularization parameter is related to the amplitude of "noise" in the input data. The input prescribed dose distribution in inverse planning does not have "noise", therefore, the non-physical oscillations in beam intensity function are primarily due to the amplification of rounding errors [9, 28]. Therefore, the value of regularization parameter is defined primarily by the beamlet functions and the optimization method. The beamlet functions and the optimization method are usually the same for a given treatment planning system, therefore, the regularization parameter is a nearly constant value for different clinical cases.

If the regularization parameter can be determined during commissioning of treatment planning system where the calculational time is not restricted, we can determine as many points on the L-curve as we need. We suppose that the L-curve is defined by a set of points $(\alpha_n, \xi_n, \eta_n)$, $n = \overline{1, N}$ and we must calculate the point of maximum curvature using this tabulated function. We note that the regularization parameter for the Tikhonov regularization is continuous and, therefore, the L-curve is a continuous smooth function (within the computational errors).

6.3.3 *Finite-Difference Approximation*

Probably, the simplest procedure to compute the curvature of the L-curve for Tikhonov regularization is to substitute the first and second derivatives in Eq. 6.23 by their finite-difference approximations:

$$\left(\xi'_\alpha\right)_n = \frac{\xi_n - \xi_{n+1}}{\alpha_n - \alpha_{n+1}}, \tag{6.24}$$

$$\left(\xi''_\alpha\right)_n = \frac{\left(\xi'_\alpha\right)_{n-1} - \left(\xi'_\alpha\right)_n}{\alpha_{n-1} - \alpha_n} \tag{6.25}$$

with similar equation for η'_α and η''_α. The main disadvantage of this approach is the generation of fluctuations that may appear because of changes in the local curvature of the L-curve. Although the L-curve is nearly smooth, the local curvature can fluctuate because of computational errors. The vertical part of the L-curve is mostly prone to these fluctuations. However, the point of maximum curvature is explicitly seen in most cases.

6.3.4 *Cubic Spline Interpolation*

Another way to compute the curvature of the L-curve is to fit a cubic spline to the discrete function $\eta_n = \eta(\xi_n), n = \overline{1, N}$ [13, 31]. Cubic spline is a cubic polynomial that is smooth in the first derivative and continuous in the second derivative within an interval and at its boundaries. Therefore, the cubic spline is a twice differentiable function, and the curvature of the L-curvature can be calculated using the formula:

$$k(\alpha) = \frac{\eta''_\xi}{\left(1 + \left(\eta'_\xi\right)^2\right)^{\frac{3}{2}}}, \tag{6.26}$$

where η'_ξ and η''_ξ are the first and second derivatives with respect to the residual norm $\xi(\alpha)$. The major disadvantage of this approach is that that the cubic spline does not have local smoothing properties. The curvature function calculated using this procedure exhibits the oscillations on the vertical part of the L-curve like the finite difference approximation.

6.3.5 *Singular Value Decomposition Analysis*

Equations 6.23 and 6.26 include both the first and second derivatives that make numerical analysis difficult. This equation can be simplified using the singular value decomposition analysis for the Tikhonov regularization. To derive a new equation for the L-curve plotted in the linear–linear scale, we introduce new variables x and y which relate to the variables ξ and η as

$$\xi = \sqrt{x}, \quad \eta = \sqrt{y}. \tag{6.27}$$

It can be shown using the singular value decomposition analysis that the following equation holds for the Tikhonov regularization [14]:

$$x'_\alpha = -\alpha y'_\alpha. \tag{6.28}$$

Differentiating Eq. 6.28 we obtain

$$x''_\alpha = -y'_\alpha - \alpha y''_\alpha. \tag{6.29}$$

Then, we calculate the first and second derivative of ξ and η using Eq. 6.28:

$$\xi'_\alpha = \frac{x'_\alpha}{2\sqrt{x}}, \quad \eta'_\alpha = \frac{y'_\alpha}{2\sqrt{y}}, \tag{6.30}$$

$$\xi''_\alpha = \frac{2xx''_\alpha - \left(x'_\alpha\right)^2}{4x^{3/2}}, \quad \eta''_\alpha = \frac{2yy''_\alpha - \left(y'_\alpha\right)^2}{4y^{3/2}}, \tag{6.31}$$

Replacing x'_α and x''_α in Eqs. 6.30 and 6.31 using Eqs. 6.28 and 6.29 and substituting the final expressions for the first and second derivatives of the residual and smoothing norms into Eq. 6.23 we obtain a new formula for the curvature of the L-curve:

$$k(\alpha) = \frac{\frac{2xy}{y'_\alpha} + \alpha x + \alpha^2 y}{\left(x + \alpha^2 y\right)^{\frac{3}{2}}}. \tag{6.32}$$

The second derivatives in Eq. 6.32 cancel each other. Equation 6.32 is much easier to compute because it contains only one first derivative y'_α with respect to the regularization parameter α.

6.3.6 Minimum of Residual and Smoothing Norms

The three previously introduced criteria are based on finding the point of maximum curvature of the L-curve. It is supposed that the point of maximum curvature is associated with the value of the regularization parameter which minimizes both residual norm $\xi(\alpha)$ and smoothing norm η(α). We can also locate the point of maximum curvature using an observation that the total of the residual and smoothing norm $\xi(\alpha) + \eta(\alpha)$ will achieve its minimum at the point of maximum curvature because the variables $\xi(\alpha)$ and $\eta(\alpha)$ achieve their minimum simultaneously [28]. We find an optimal value of the regularization parameter α_{opt} by plotting the total sum $\xi(\alpha) + \eta(\alpha)$:

$$\alpha_{opt} = \arg\min\{\xi(\alpha) + \eta(\alpha)\}, \tag{6.33}$$

that achieves its minimum at the point of maximum curvature. An advantage if this method is that that it does not include calculation of any derivatives. However, the method requires a detailed dependence of both norms $\eta(\alpha)$ and $\xi(\alpha)$ on the regularization parameter α. This is not a problem because, as we already mentioned, we can calculate as many points as we need to simulate the tabulated function $(\alpha_n, \xi_n, \eta_n), n = \overline{1, N}$.

6.4 Model Problems for Radiotherapy Optimization

6.4.1 *Inverse Planning Algorithm*

We consider an inverse treatment planning algorithm based the L-BFGS-B optimization code that was developed at the Optimization Technology Center, a joint venture of Argonne National Laboratory and Northwestern University [22]. To find a minimum of the objective function given by Eq. 6.7, the optimization code uses the L-BFGS-B algorithm that is a limited-memory version of the quasi-Newton method with simple constraints on the variables. The quasi-Newton method requires an evaluation of the objective function and only its gradients at every iteration. The first order Tikhonov stabilizing functional usually provides the optimal combination of smoothing and approximation for inverse treatment planning [7, 8]; therefore, we utilize the first order stabilizing functional which is given by Eq. 6.18. The different numerical methods are applied to calculate the point of maximum curvature of the L-curves, that is a criterion to locate the corner.

To find the minimum of modified objective function given by Eq. 6.7 using the quasi-Newton algorithm, we need to calculate the objective function and its gradients at each iteration. The modified objective function includes the quadratic objective function and the stabilizing function. The discrete approximations for quadratic objective function given by Eq. 6.8 and its gradients are streiworward [9];therefore, we describe here only approximations for the stabilizing terms and their gradients.

The beam intensity function $f(\xi)$ is a two-dimensional function for a general three-dimensional case. We will describe a two-directional case which can be easily generalized to a three-dimensional case. If the intensity function $f(\xi)$ is defined by its weights $w_m, m = \overline{1, M}$, the following discretization can be applied for the calculation of intensity function $f(\xi)$ and its first derivative:

$$f_m = \frac{w_m}{\Delta_m}, \tag{6.34}$$

$$\left(f'_\xi\right)_m = \frac{w_m - w_{m+1}}{\Delta_m^2}, \tag{6.35}$$

where Δ_m is the width of the m-th interval. The first order stabilizing functions defined can be written as

$$\Omega_1 = \sum_{m=1}^{M} \left\{ \frac{w_m^2}{\Delta_m} + \frac{(w_m - w_{m+1})^2}{\Delta_m^3}. \right\} \tag{6.36}$$

To conduct the iterative search for the minimum of objective function we need the derivatives of the stabilizing functions with respect to the weights w_m, $m = \overline{1, M}$. Differentiating Eq. 6.36 with respect to the weight w_m we obtain the gradients of the stabilizing functions:

$$\frac{d\Omega_1}{dw_m} = \frac{2w_m}{\Delta_m} + \frac{2(2w_m - w_{m+1} - w_{m-1})}{\Delta_m^3}, \quad m = \overline{1, M}. \tag{6.37}$$

6.4.2 Beamlet Dose Functions

The beamlet dose function $d_{\mathrm{m}}(\boldsymbol{r})$ in Eq. 6.2 was simulated as the dose distribution in water from a narrow beam with square collimation normally incident upon a semi-infinite water phantom. The calculations have been performed in orthogonal Cartesian coordinate system using the EGSnrc/DOSXYZnrc Monte Carlo code [32, 33]. In the 3D simulation, the central axis of the beamlet coincides with the Z axis and X and Y are transversal coordinates. The beamlet dose distributions for 2D optimization were obtained by integration of the 3D dose distribution over one of the transversal coordinates (Y). The beamlet dose distributions distribution from the square collimation of 0.3 cm × 0.3 cm of incident beam calculated for a typical 6 MV photon spectrum from a linear accelerator using 5.0E+8 Monte Carlo histories is shown in Fig. 6.1.

To consider the scattered radiation correctly, we calculated the 3D lateral beamlet transport up to distance of 10 cm in each of the transversal X and Y coordinates. The importance of scattered radiation from a narrow beamlet in the IMRT dose calculations was demonstrated [34]. We have used variable grid size to optimize the computational time. The grid size changed from 0.05 cm in the steep dose gradient region to 0.5 cm in the regions with slow changing solution.

6.4.3 Radiotherapy Model Problem

To analyze and test the performance of the L-curve method, we apply the variational regularization technique to a typical radiotherapy treatment model problem with intensity-modulated beams in head-and-neck cancer shown in Fig. 6.2. The problem represents a C-shaped planning target volume (PTV) wrapping around the spinal cord. The 2D phantom is approximated on a 0.3 cm × 0.3 cm computational grid. The phantom is irradiated with nine co-planar 6 MV photon beams uniformly distributed

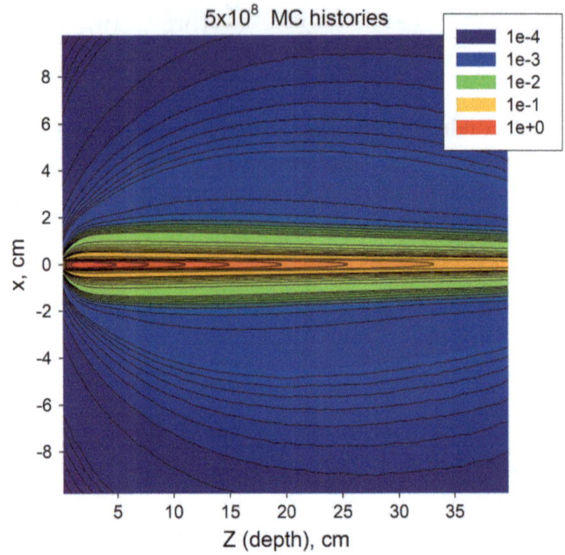

Fig. 6.1 Dose distribution from the 6 MV 3 mm beamlet calculated using Monte Carlo method with 5 $\times$ 10^8 Monte Carlo histories. Isodose lines are plotted using logarithmic scale

between 0° and 360°. The prescription dose for the PTV and normal tissue was selected to be 100% and 0% respectively and the importance factors have been set to 100 for the PTV and 1 for the normal tissue. The prescription dose and importance factor for the critical structures (parotid glands and cord) have been selected as 10% and 10.

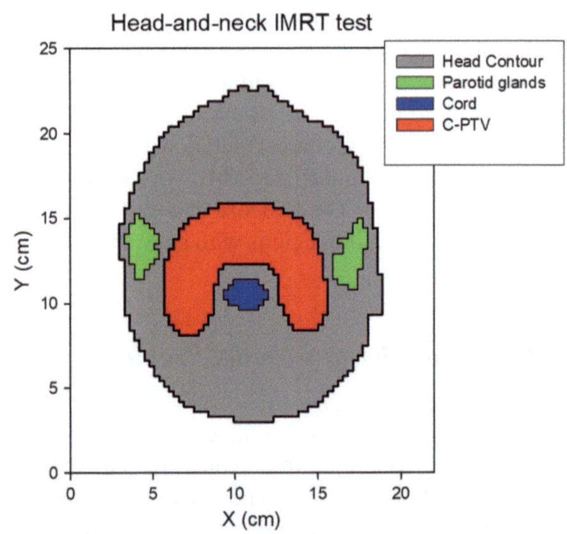

Fig. 6.2 Two-dimensional phantom in radiotherapy model for head and neck cancer

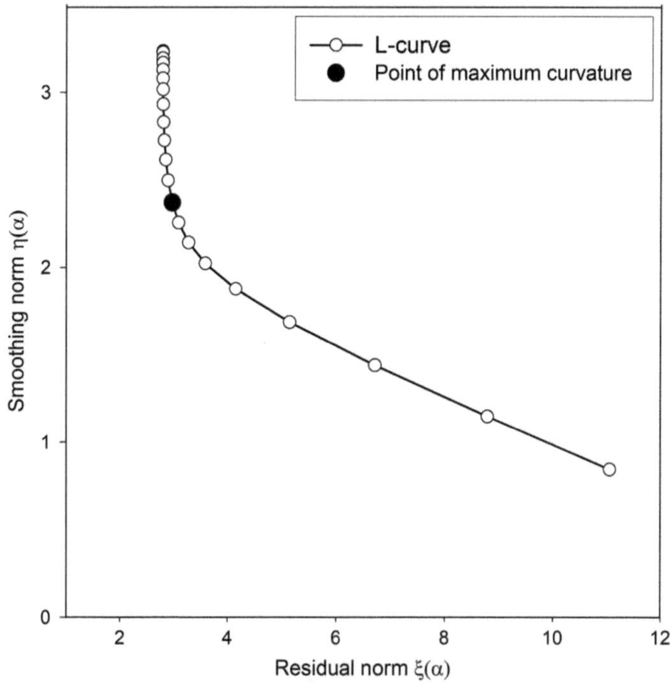

Fig. 6.3 L-curve for radiotherapy model problem in head and neck cancer

The L-curve for the test problem is presented in Fig. 6.3. The characteristic curve exhibits the L-shape with the vertical and horizontal parts. We note that it is recommended to plot the L-curve in the log–log scale to emphasize the "flat" parts and the corner. Our numerical results indicate, however, that the L-shape for inverse planning is better seen in the linear–linear scale. This may be attributed to the fact that inverse treatment planning has "noise" free input data (prescribed dose distribution).

The curvature functions of the L-curve presented in Fig. 6.3 are shown in Fig. 6.4a–c. These curvature functions were calculated using finite-difference approximation, cubic spline interpolation and the formula based on the singular value decomposition analysis. We show that the point of maximum curvature can be determined in a most robust way using a formula derived from the singular value decomposition analysis. We see that all three methods can predict the curvature of the L-curve for the horizontal part. The finite-difference approximation and the cubic spline interpolation that include calculation of both first and second derivatives fail on the vertical part of the L-curve. White the maximum in the curvature function can be located for all methods, the largest numerical fluctuations are obtained for the cubic spline interpolation. Only the formula based on the singular value decomposition is robust on both vertical and horizontal parts. We can conclude that the formula based on the singular value decomposition analysis is a method of choice for locating the point of maximum curvature on the L-curve.

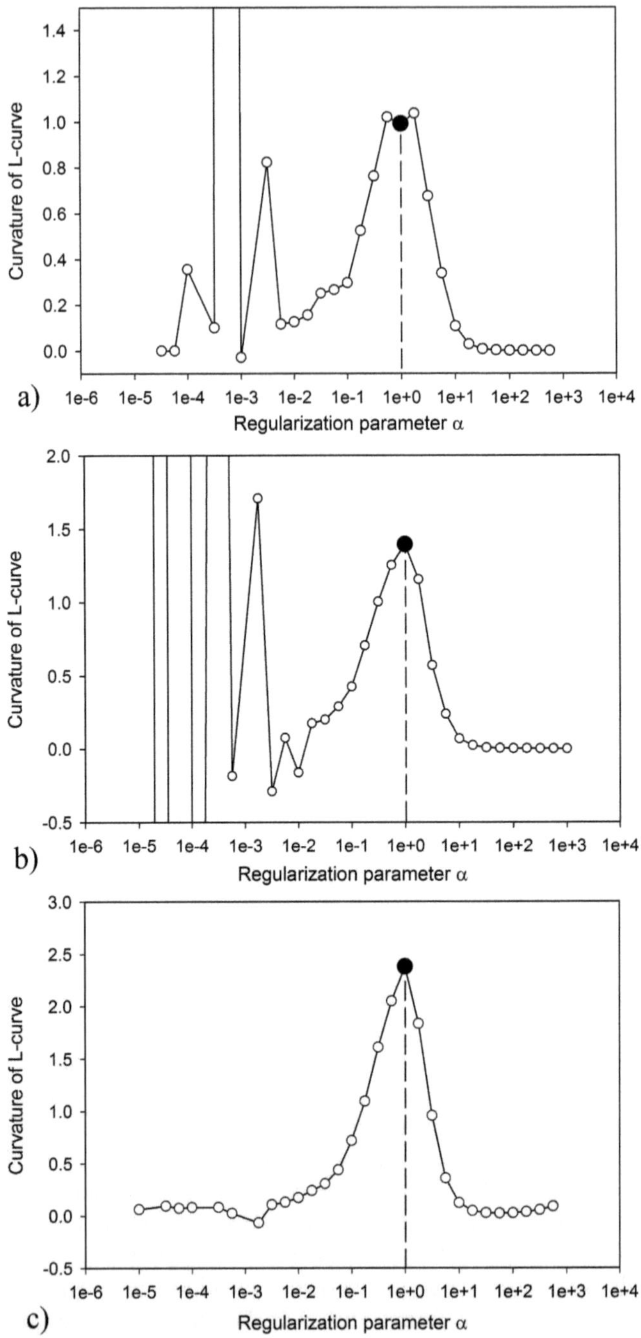

Fig. 6.4 Curvature of the L-curve for radiotherapy model problem in head and neck cancer calculated by **a** finite-difference approximation; **b** cubic spline interpolation; and **c** singular value decomposition analysis [28]

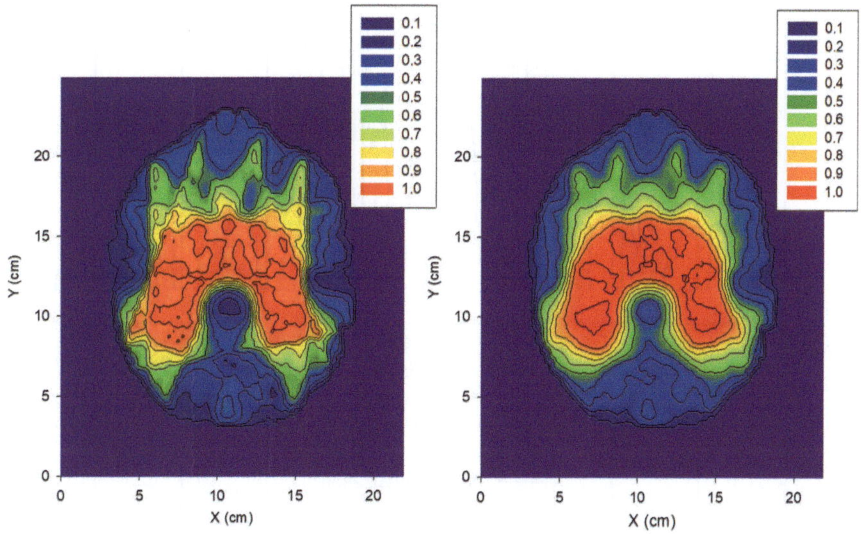

Fig. 6.5 Isodose distributions calculated for radiotherapy model problem in head and neck cancer: **a** not regularized and **b** regularized

In Figs. 6.5a, b and 6.6a, b, we show isodose distributions and beam intensity functions calculated with regularization and without regularization techniques. The regularized beam intensity functions and dose distributions are the smooth functions that are more practical for dose prescription and delivery. From the properties of the L-curve we know that this improvement in beam intensity functions was achieved practically without deterioration of fit to the prescribed dose distribution which is described by residual norm.

We have applied to inverse treatment planning the L-curve method that is an important graphical tool for selection of the regularization parameter in the discrete ill-posed problems. For typical models of prostate cancer treatment, we have obtained the distinctive L-shaped dependence of the smoothing norm of the solution from the residual norm. We have applied different numerical methods to compute the "corner" of the L-curves where the optimal value of regularization parameter is found. We found that the formula based on the singular value decomposition analysis is the most robust tool for locating the corner of the L-curve. The regularized beam intensity functions and the dose distributions have practically acceptable smoothness and accuracy when the "corner" value of the regularization parameter is applied. The general conclusion is that that the Tikhonov regularization supplemented by the L-curve method for determining the regularization parameter is a complete method that can be applied to practical problems of inverse treatment planning.

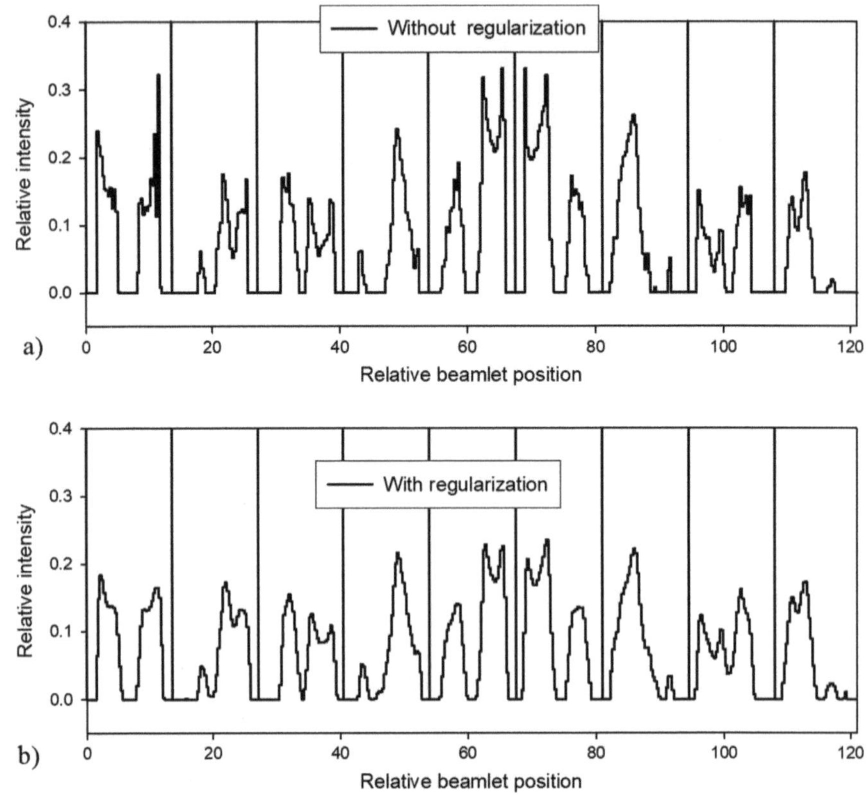

Fig. 6.6 Beam intensity functions for radiotherapy model problem in head and neck cancer computed without **a** and with **b** variational regularization

6.5 EUD-Based Dose Optimization

6.5.1 Advantages and Difficulties in EUD Optimization

As we already mentioned in Chap. 4, the dose distributions in radiation therapy are usually inhomogeneous in the target volumes and normal tissues. Each treatment plan has a certain pattern of dose non-uniformity for equivalent dose prescription that produces a certain individual radiobiological response. This complicates the treatment outcome analysis and comparison of different treatment plans. The direct optimization of the EUD in inverse planning for IMRT prevents variation of radio-biological effect between different radiotherapy treatment plans that may appear due to variation in the pattern of dose non-uniformity [35–38]. For instance, the survival fraction of clonogenic cells would be consistent with the prescription when the optimized EUD is equal to the prescribed EUD.

One of the problems in the practical implementation of this approach is that the spatial dose distribution in the EUD-based inverse planning would be underdetermined because an unlimited number of nonuniform dose distributions can be computed for a prescribed value of EUD. The EUD is a functional of the spatial dose distribution function that does not maintain the information about this function. There are many non-uniform dose distributions that produce the same EUD value. Therefore, one of the problems in EUD-based inverse planning is that the spatial dose distribution is underdetermined in these calculations.

Together with the ill-posedness of the underlying integral equation, this effect may significantly increase the dose non-uniformity in inverse treatment planning. To optimize the EUD and keep dose non-uniformity within reasonable limits, we implemented into the EUD-based objective function an additional criterion that ensures the smoothness of beam intensity functions. This approach is like the variational regularization technique that was previously studied for the dose-based least squares optimization. We show that the variational regularization together with the L-curve criterion for the regularization parameter can significantly reduce dose nonuniformity in EUD-based inverse planning.

We discuss the EUD in objective functions for IMRT optimization and propose a version of EUD-based IMRT optimization model that explicitly controls the uniformity of the dose distribution. Assuming that the optimization problem is solved accurately and the ill-posedness of the underlying Fredholm integral equation of the first kind is not considered, we can see that there are many nonuniform dose distributions (some of them are not clinically applicable) which meet an EUD-based objective. In combination with the ill-posedness of the Fredholm integral equation of the first kind which governs the optimization process, this can produce strongly non-uniform dose distributions. However, it is usually assumed that a smooth dose distribution without significant hot and cold spots is clinically preferable provided it produces the same EUD. Therefore, this problem can be overcome by introducing additional knowledge about desirable solutions into the optimization process. For instance, this can be done using a hybrid dose-based/EUD-based optimization [39]

6.5.2 EUD-Based Objective Functions

Treatment planning algorithms usually utilize the generalized concept of equivalent uniform dose (gEUD) that is given by

$$EUD = \left\{ \frac{1}{N} \sum_{n=1}^{N} D_n^{\alpha} \right\}^{1/\alpha}, \tag{6.38}$$

where D_{n} is the dose in the voxel n, N is the total number of voxels and a is the parameter defined by the radiobiological model [36]. In the inverse planning algorithms, the voxel dose D_{m} is usually calculated using the beamlet model that presents the discretized form of the Fredholm integral equation of the first kind. Equation 6.38 is convenient for use in optimization algorithms because it is based only on one adjustable radiobiological parameter a, and it can be applied to tumors and normal tissues. The parameter a is negative for tumors and positive for normal tissues [36–40].

We consider an objective function for the EUD-based inverse planning that preserves the prescribed value of the EUD and controls dose nonuniformity. To achieve this goal, we have generalized for the EUD-based optimization the variational regularization technique which was studied for the dose-based optimization with least squares objective function. Stability in these calculations was achieved using a special smoothing norm in the least squares objective function [9, 40]. It can be theoretically shown that the implementation of this norm into least squares objective function makes the optimization problem well-posed. Numerical simulations show that variational regularization can improve optimization for more complex non-linear objective functions, for example, the optimization of the EUD.

To study optimization of the EUD, we use the following objective function with a regularization term:

$$\boldsymbol{w}_\alpha = \arg\min\left\{\sum_{k=1}^{K} I_k\left(EUD_k(\boldsymbol{w}_\alpha) - EUD_k^{pr}\right)^2 + \alpha\Omega(\boldsymbol{w}_\alpha)\right\}. \tag{6.39}$$

where $\boldsymbol{w}_\alpha = (w_m, m = \overline{1, M})$ are the beamlet weights, EUD_k^{pr} and $EUD_k(\boldsymbol{w}_\alpha)$ are the prescribed and calculated EUD for the structure k respectively, and I_k is the importance factor. The spatial dose distribution in the EUD based optimization problem given by Eq. 6.39 is underdetermined for the regularization parameter $\alpha = 0$. In this case, there is an unlimited number of nonuniform dose distributions (and correspondingly beam intensity functions) that satisfy the EUD-based objective function. Together with the ill-posedness of the underlying integral Eq. 6.1, this may increase the non-uniformity in the optimized dose distributions. To achieve stability of calculations, we modified the optimization problem with a stabilizing functional Ω multiplied by a regularization parameter α [9, 40]. The stabilizing functional Ω is calculated as a smoothing norm of the beam intensity function; therefore, we restrict the class of solutions by the smoothest solution. Based on our previous results for dose-based inverse treatment planning [9], we utilize the first order smoothing norm given by Eq. 6.35.

The objective function given by Eq. 6.39 cannot be used effectively without a method for the selection of the optimal value of regularization parameter α. We demonstrate numerically that the L-curve criterion for regularization parameter α applied previously to the least squares dose-based optimization can be effectively applied to the EUD-based optimization. The main idea of the L-curve method is to plot the smoothing norm $\eta(\alpha)$ as a function of the residual norm $\xi(\alpha)$ for all values of the regularization parameter α [13]. The smoothing norm $\eta(\alpha)$ and the residual norm $\xi(\alpha)$ for EUD based optimization can be derived similar to dose-based optimization in Eqs. 6.21 and 6.22. The function has an L-shaped form with distinctive vertical and horizontal parts. The residual and smoothing norms simultaneously achieve their minimum at the corner; therefore, the optimal value of the regularization parameter is found in the vicinity of corner. The corner can be located, for instance, using the criterion of maximum curvature of the L-curve. Using the L-curve method, we can find the smoothest solution which meets the EUD-based objective function.

6.5.3 Model Problem for EUD-Based Optimization

To study the EUD-based optimization problems, we utilize the research inverse treatment planning system that was previously described for the dose-based optimization. This inverse treatment planning system is based on Monte Carlo pre-calculated beamlet dose functions and a quasi-Newton optimization algorithm. The inverse treatment planning system was modified to include the EUD-based objective function and its gradients.

We applied the EUD-based optimization algorithm to a model case of head and neck cancer treatment optimization shown in Fig. 6.2. The head and neck phantom is irradiated with nine 6 MV beams equally placed between the angles 0° and 360°. Critical structures include left and right parotid glands and spinal cord. The EUD prescription was defined as 1.0 for target volumes, 0.1 for critical structures and 0.0 for normal tissue. The importance factors were set as 1.0 for target volumes, 0.1 for critical structures and 0.01 for normal tissue. The value of radiobiological parameter was set $a = -10$ for targets, $a = 5$ for critical structures and $a = 10$ for normal tissue. The value of radiobiological parameter a did not affect the convergence properties of the optimization algorithm. We also note that the computational time was significantly reduced in the EUD optimization with regularization because the total number of iterations was reduced by a factor of approximately 4.

We used the L-curve method to obtain the optimal value of the regularization parameter α. In Fig. 6.7a, we show the L-curve dependence of the smoothing norm as a function of the residual norm for the EUD-based optimization in the head-and-neck model problems. The curvature of the L-curves is computed using the finite-difference approximation described by Eqs. 6.24 and 6.25. The optimal value of the regularization parameter is selected at the point of maximum curvature corresponding to the corner of the L-curve. The regularization parameter is needed to suppress the numerical fluctuations that depend primarily on the input data and truncation errors in the numerical algorithms used in solving the optimization model.

In Fig. 6.8a, b, we show the dose distributions obtained for head and neck cancer model problem using the EUD-based objective function with and without the regularization term. The dose distribution obtained without regularization term is strongly inhomogeneous and exhibits significant hot spots. The value of hot spots is almost double the value of prescribed EUD (1.79 for the prescribed EUD = 1.0). The regularized plan is characterized by smoother iso-dose distributions and better uniformity of dose distribution within the planning target volume. We note that the spatial dose

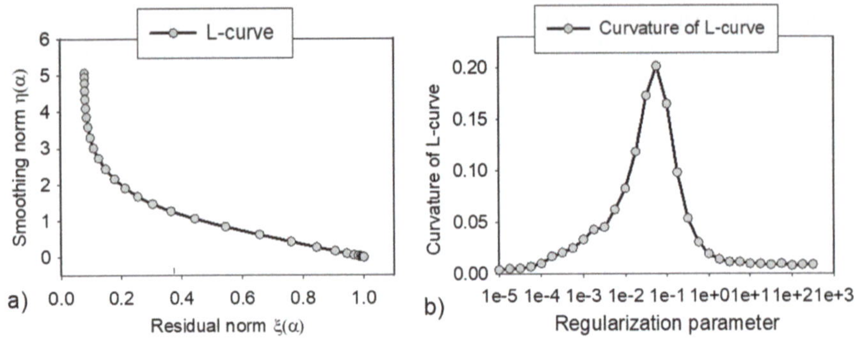

Fig. 6.7 L-curve **a** and its curvature **b** in radiotherapy problem for head and neck cancer with EUD optimization.

distributions shown in Fig. 6.8a, b are different, but they have the same value of EUD in the target volumes.

In Fig. 6.9, we show the DVHs for PTV calculated for head and neck model problem using different methods. The DVHs obtained using the dose-based optimization are shown for comparison. The DVHs for critical structures are shown in Fig. 6.10. The following conclusions are valid regarding the EUD-based optimization without regularization. First, the DVHs for the target volumes have smoother gradients in the EUD-based optimization compared to the dose-based optimization

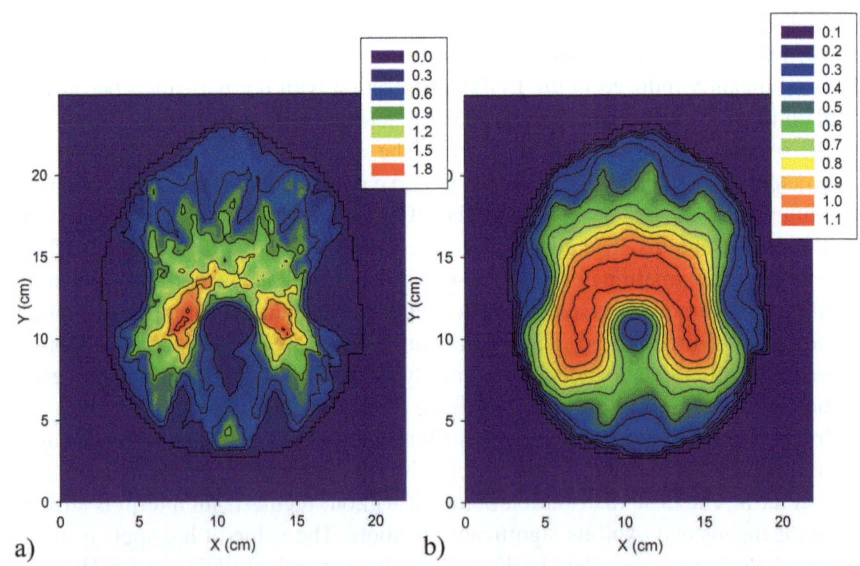

Fig. 6.8 Dose distributions in EUD-based optimization in radiotherapy for head and neck cancer: **a** EUD objective function; **b** EUD objective function with regularization [40]

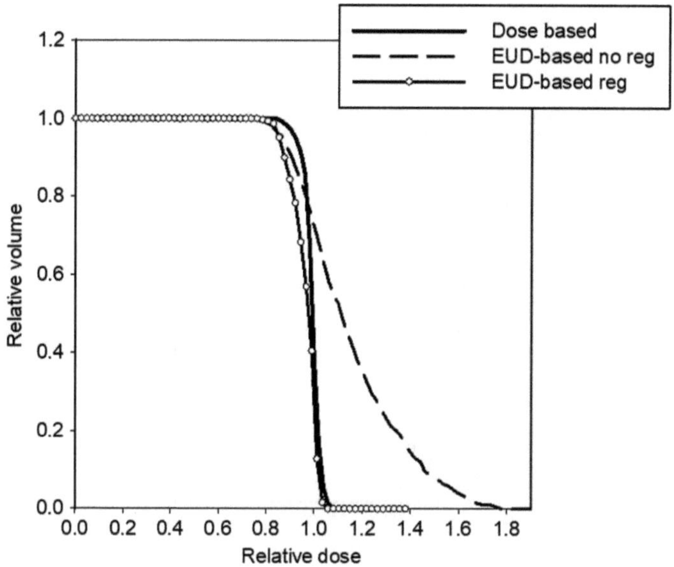

Fig. 6.9 Dose-volume histograms for PTV in radiotherapy problem for head and neck cancer

because of large hot spots. Second, the EUD-based optimization without regularization significantly reduces the dose received by critical structures. Similar results for the target volumes and critical structures have been previously obtained [37].

By adding a regularization method to the EUD-based optimization, the following effects can be seen. Firstly, the dose to critical structures was increased, however, it stayed within the practical limits and was found to be comparable to the level obtained with dose-based optimization. Second, the DVHs for target volumes approach the curves obtained with the dose-based optimization, however, we were not able to obtain the same high gradient values. The results for the target volumes can be partially explained by the fact that the dose nonuniformity in EUD-based optimization is larger than in dose-based optimization even if the strongest regularization is applied. Also, smoother DVH gradient can be explained by the smoother dose gradients between structures which are due to the regularization procedure. Apparently, it is not possible to compute numerically the same DVH gradient for targets when optimizing only average values as EUD. However, the EUD-based optimization provides better biological conformation at the cost of deterioration of physical conformation. Under biological conformation we assume keeping the same fraction of surviving clonogenic cells in the target volume.

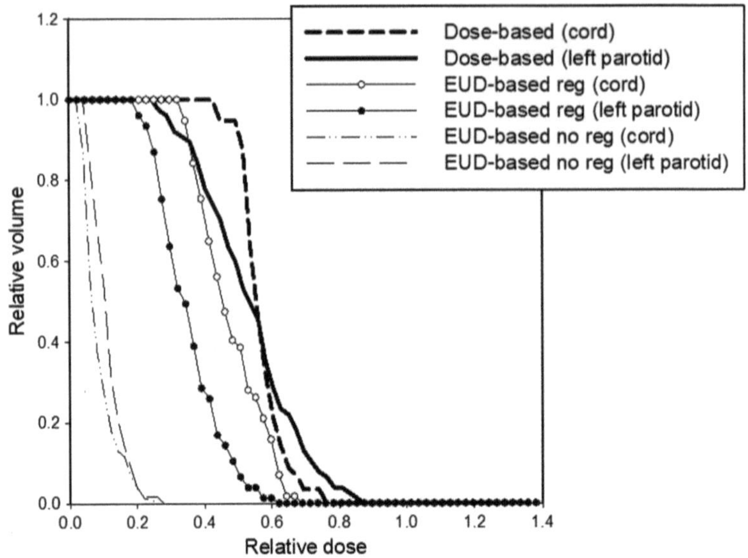

Fig. 6.10 Dose-volume histograms for critical structures in head and neck radiotherapy model problem [40]

6.6 Summary of EUD-Based Optimization

Dose distributions produced by the EUD-based treatment planning optimization are typically nonuniform with significant hot spots in the target volumes. To overcome this difficulty, we have generalized the variational regularization method for solving ill-posed problems for EUD-based IMRT optimization. We have applied to EUD-based optimization the variational regularization technique with the L-curve method for determining an optimal value of the regularization parameter. The regularization parameter is found using the L-curve method that is based on minimization of the residual norm that is a measure of accuracy of fit and the smoothing norm that a measure of smoothness of beam intensity functions. As a result, we find the smoothest dose distribution for a prescribed value of EUD. We demonstrate that the regularization method decreases mean dose and standard deviation in the target volume while keeping an EUD of the same value. Implementation of the variational regularization technique into an EUD-based IMRT treatment optimization allows one to find optimal balance between robustness of solutions and optimization accuracy.

One of main thus advantages of regularized EUD-based optimization is that the optimized beam intensity functions are smooth thus improving the dose delivery. Post-processing by filtering to obtain smoother distributions is not desirable in EUD optimization. This procedure changes the pattern of dose nonuniformity and correspondingly the value of EUD. It is assumed that there is an uncertainty in EUD associated with dose prescription in dose-based optimization; therefore, the post-processing is acceptable. In the EUD-based optimization, we need to maintain the same optimized value of the EUD. Therefore, the EUD-based optimization algorithm should produce the beam fluence functions that are suitable for delivery and post-processing is not desirable.

6.7 Optimization of EUAD for Hypoxic Tumors

6.7.1 Two-Step Optimization of EUAD

In this Chapter, we described in detail mathematical algorithms for the dose-based optimization and the EUD based optimization in radiotherapy treatment planning. One of these approaches can be applied to optimization of the EUAD depending on the specific problem. The EUAD in heterogeneous hypoxic tumors can be used for patient-specific biological optimization because maximization of the EUAD allows one to find the optimal non-uniform dose distributions. The application of the EUAD to dose optimization in hypoxic tumors is a robust alternative to the TCP optimization because the EUAD does not require the density of clonogenic cells. At the same time, the EUAD concept allows one to consider intra-tumor radiosensitivity and volume effects.

The dose-based optimization is a robust approach for the EUAD optimization, and this radiotherapy optimization problem should be solved in two steps. First, the non-uniform dose distribution $D_\alpha(\boldsymbol{r})$ should be found from minimization of the following objective function:

$$D_\alpha(\boldsymbol{r}) = \arg\min\left\{\sum_{k=1}^{K}\left(EUAD_k(D_\alpha(\boldsymbol{r}), OER(\boldsymbol{r})) - EUAD_k^{pr}\right)^2\right\}, \tag{6.40}$$

where EUAD_k^{pr} is prescribed EUAD in sub-volume k, $D_\alpha(\boldsymbol{r})$ is the non-uniform distribution of physical dose, and $\mathrm{OER}(\boldsymbol{r})$ is the distribution of OER. The optimization problem given by Eq. 6.40 does not require regularization because it does not involve the Fredholm integral equation of the first kind described by Eqs. 2.20 and 6.1.

When the optimal distribution of physical dose $D_\alpha(\boldsymbol{r})$ is found, it can be used in Eq. 6.7 to find the beam intensity functions. This second step of optimization will require regularization with the regularization parameter that can be found from the L-curves.

6.7.2 Direct Optimization of EUAD

An alternative approach to the two-step optimization of the EUAD, the direct optimization of the EUAD can be used in a way that is similar to the EUD optimization problem described by Eq. (6.39). We discussed the advantages of direct EUD optimization and similar optimization algorithms can be applied to the EUAD optimization. The optimized beam profiles for the EUAD-based objective function can be found solving the following optimization problem:

$$\boldsymbol{w}_{\alpha} = \arg\min\left\{\sum_{k=1}^{K} I_k\left(EUAD_k(\boldsymbol{w}_{\boldsymbol{\alpha}}) - EUAD_k^{pr}\right)^2 + \alpha\Omega(\boldsymbol{w}_{\alpha})\right\}, \quad (6.41)$$

$\boldsymbol{w}_{\boldsymbol{\alpha}} = (w_m, m = \overline{1, M})$ are the beamlet weights, $\text{EUAD}_k(w_\alpha)$ and EUAD_m^{pr} and are the calculated and prescribed EUAD for the structure k, and I_k is the importance factor. The EUAD optimization problem described by Eq. 6.41 has advantages and disadvantages like the EUD optimization problem described by Eq. 6.39. Specifically, this approach may produce significant instabilities in the reconstructed beam profiles that require regularization.

We note that problems like those discussed for EUD and EUAD optimization may appear in treatment plan optimization with more complicated objective functions based on TCP models [1–4]. Spatial dose distribution is also underdetermined in these TCP optimization problems; therefore, dose nonuniformity may increase compared to optimization with the dose-based objectives. The smoothing algorithm presented in this article is universal and can be applied to stabilize the TCP optimization as well.

6.7.3 EUAD Optimization and Alternative Approaches to Treat Hypoxia

The theoretical results presented in this book support the idea that the dose distributions in cancer radiation therapy can be optimized based on the hypoxia images. However, hypoxia-targeted dose escalation has not yet found a widespread application in the clinic. Many research studies on functional image-guided dose escalation have been treatment planning studies, and treatment outcome have not been studied. Challenges for the clinical application of hypoxia targeted dose escalation include the risks associated with normal tissue complications and dynamic nature of the location and extent of hypoxia. For example, temporal variation of hypoxic regions, target deformation and tumor motion can generate delivery errors that would compromise precise targeting of hypoxia using non-uniform dose distribution.

We reviewed the mathematical aspects and optimization of radiotherapy response for hypoxic tumors. We introduced and tested the parsimonious models of radiotherapy response and tested them using the serial imaging data acquired during fractionated radiotherapy. These tumor response models were used for analysis of the TCP and development of new predictive parameters in hypoxic tumors such as the EUAD. We discussed the derivation of the OER from FMISO PET images and using it in advanced parameters such as the EUAD. We have demonstrated the predictive value of the EUAD in radiotherapy for glioblastoma.

Using the concept of the EUAD we introduced a new parameter called the DNE and have shown that the nonuniform dose distributions can produce higher TCP in heterogeneous hypoxic tumors. However, this effect is observed only of the OER is larger than a threshold value OER_{th}. Our simulations show that volume dependence in the TCP is determined predominantly by hypoxic volume. Therefore, the TCP reduction can be described largely by increasing the hypoxic tumor volume with the total tumor volume.

Our simulations show that hypoxia-targeted dose escalation is biologically more efficient for hypofractionated dose regimens; therefore, widespread implementation of hypofractionated dose regimens may stimulate wider interest in the hypoxia-targeted dose escalation. Dose de-escalation based on hypoxia measurements is also investigated [41]. The clinical implementation of radiotherapy treatment machines that integrate medical linear accelerators with PET or MRI could increase the application of hypoxia imaging in radiation oncology and potentially improve treatment outcomes [42]. Alternative methods to target hypoxia may include application of high-LET particles and radio sensitization [43].

References

1. Sövik A, Malinen E, Olsen D. Strategies for biologic image guided dose escalation: a review. Int J Radiat Oncol Biol Phys. 2009, 73: 650–658.
2. Yang Y, Xing L, Towards biologically conformal radiation therapy (BCRT): Selective IMRT dose escalation under the guidance of spatial biology distribution, Med Phys. 2005; 32 (6): 1473–1484.
3. Malinen E, Sovik A, Hristov D, Bruland ØS, Olsen DR. Adapting radiotherapy to hypoxic tumors. Phys Med Biol. 2006; 51: 4903–4921.
4. Thorwarth D, Eschmann SM, Paulsen F, Alber M. Hypoxia dose painting by numbers: A planning study. Int J Radiat Oncol Biol Phys. 2007; 68 (1):291–300.
5. Webb S, Optimizing the planning of intensity-modulated radiotherapy. Phys Med Biol. 1994; 39: 2229–2246.
6. Bortfeld T, Bürkelbach J, Boesecke R, Schlegel W. Methods of image reconstruction from projections applied to conformation radiotherapy. Phys Med Biol. 1990; 35: 1423–1434.
7. Phillips DL. A technique for the numerical solution of certain integral equations of the first kind. Journal of the Association for Computing Machinery. 1962; 9: 84–97.
8. Tikhonov AN. Regularization of incorrectly posed problems. Soviet Mathematics, Translation of Doklady Akademii Nauk SSSR, Vol. 4, pp. 1624–1627 (American Mathematical Society, Providence, Rhode Island, 1963).
9. Chvetsov AV, Calvetti D, Sohn JW, Kinsella TJ. Regularization of inverse treatment planning for intensity modulated radiation therapy. Med Phys. 2005; **32**: 501–514.

10. Webb S, Convery DJ, Evans PM. Inverse planning with constraints to generate smoothed intensity-modulated beams. Phys Med Biol. 1998; 43, 2785–2794.
11. Mohan R, Arnfield M, Tong S, Wu Q, Siebers J. The impact of fluctuations in intensity patterns on the number of monitor units and the quality and accuracy of intensity modulated radiotherapy. Med Phys. 2000; **27**: 1226–1237.
12. Wang JZ, Li XA, D'Sousa WD, Stewart RD. Impact of prolonged fractionation delivery times on tumor control: A note of caution for intensity-modulated radiation therapy (IMRT). Int J Radiat Oncol Biol Phys. 2003; 57: 543–552.
13. Hansen PC. Rank-Deficient and Discrete Ill-Posed Problems. Philadelphia, PA, SIAM, 1998.
14. Vogel CR. Computational Methods for Inverse Problems. Philadelphia, PA, SIAM, 2002.
15. Brahme A. Optimization of stationary and moving beam radiation therapy techniques. Radiother Oncol. 1988; 12: 129–140.
16. Censor Y, Altschuler MD, Powlis WD. On the use of Cimmino's simultaneous projections method for computing a solution of the inverse problem in radiation therapy treatment planning. Inverse Problems 1988; 4: 607–623.
17. Xiao Y, Michalski D, Censor Y, Galvin JM. Inherent smoothness of intensity patterns for intensity modulated radiation therapy generated by simultaneous projection algorithm. Phys Med Biol. 2004; 49: 3227–3245.
18. Michalski D, Xiao Y, Censor Y, Galvin JM. The dose-volume constraint satisfaction problem for inverse treatment planning with field segments. Phys Med Biol. 2004; 49: 601–616.
19. Wu C, Jeraj R, Lu W, Mackie TW. Fast treatment plan modification with an over-relaxed Cimmino algorithm. Med Phys. 2004; 31:191–200.
20. Alber M, Nüsslin F. Intensity modulated photon beams subject to a minimal surface smoothing constraint. Phys Med Biol. 2000; 45: N49–N52.
21. Spirou SV, Fournier-Bidoz N, Yang J, Chui CS, Ling CC, Smoothing intensity modulated beam profiles to improve the efficiency of delivery. Med. Phys. 2001; 28: 2105–2112.
22. Byrd R H, Lu P, Nocedal J, Zhu C. A limited memory algorithm for bound constrained optimization. SIAM J Sci Comput. 1995; 16:1190–1208.
23. Corana A, Marchesi M, Martini C, Ridella S. Minimizing multimodal functions of continuous variables with the "simulated annealing" algorithm. ACM Trans Math Software. 1987; 13(3): 262–280.
24. Goffe WL, Ferrier GD, Rogers J. Global optimization of statistical functions with simulated annealing. J Econometrics 1994; 60(1/2): 65–99.
25. Hansen PC. Analysis of discrete ill-posed problems by means of the L-curve. SIAM Review. 1992; 34(4): 561–580.
26. Wei J, Sandison GA, Chvetsov AV. Reconstruction of electron spectra from depth doses with adaptive regularization. Med Phys. 2006; 33(2): 354–359.
27. Sourbron S, Luypaert R, Van Schuerbeek P, Dujardin M, Stadnik T. Choice of the regularization parameter for perfusion quantification with MRI. Phys Med Biol. 2004; 49: 3307–3324.
28. Chvetsov AV. L-curve analysis of radiotherapy optimization problems Med Phys. 2005; 32(8): 2598–2605.
29. Craft D, Süss P, Bortfeld T. The tradeoff between treatment plan quality and required number of monitor units in intensity-modulated radiotherapy Int J. Radiat Oncol Biol Phys. 2007; 67: 1596–1605.
30. Calvetti D, Golub GH, Reichel L. Estimation of the L-curve via Lanczos bidiagonalization. BIT. 1999; 39, 603–619.
31. Press WH, Vetterling WT, Teukolsky SA, Flannery BP. Numerical Recipes in Fortran 77: The Art of Scientific Computing, Cambridge, Cambridge University Press, 1992.
32. Kawrakow I, Rogers DWO, The EGSnrc Code System: Monte Carlo simulation of electron and photon transport, Technical Report PIRS-701. Ottawa, Canada, National Research Council of Canada, 2000.
33. Walters BRB, Rogers DWO, DOSXYZnrc User's Manual, Technical Report PIRS-794, Ottawa, Canada, National Research Council of Canada, 2002.

34. Zakarian C, Deasy JO. Beamlet dose distribution compression and reconstruction using wavelets for intensity modulated treatment planning. Med Phys. 2004; 31, 368–375.
35. Niemierko A. Reporting and analyzing dose distributions: A concept of equivalent uniform dose. Med Phys. 1997; 24: 103-110.
36. Niemierko A. A generalized concept of Equivalent Uniform Dose (EUD). Med Phys. 1999; 28: 1100 (abstract).
37. Wu Q, Mohan R, Niemierko A, Schmidt-Ullrich R. Optimization of intensity-modulated radiotherapy plans based on the equivalent uniform dose. Int J Radiat Oncol Biol Phys. 2002; 52: 224–235.
38. Choi B, Deasy JO. The generalized equivalent uniform dose function as a basis for intensity-modulated treatment planning Phys Med Biol. 2002; 47: 3579–3589.
39. Thieke C, Bortfeld T, Niemierko A, Nill S. From physical dose constraints to equivalent dose constraints in inverse treatment planning. Med Phys. 2003; 30(9): 2332–2339.
40. Chvetsov AV, Dempsey JF, Palta JR. Optimization of equivalent uniform dose using the L-curve criterion. Phys Med Biol. 2007; 52: 5973–5984.
41. Lee N, Schoder H, Beattie B, Lanning R, Riaz N, McBride S, Katabi N, Li D, Yarusi B, Chan S, Mitrani L, Zhang Z, Pfister DG, Sherman E, Baxi S, Boyle J, Morris LG, Ganly I, Wong R, Humm J. Strategy of Using Intratreatment Hypoxia Imaging to Selectively and Safely Guide Radiation Dose De-escalation Concurrent With Chemotherapy for Locoregionally Advanced Human Papillomaviruse - Related Oropharyngeal Carcinoma, Int J Radiation Oncol Biol Phys. 2016; 96(1): 9–17.
42. Mirzakhanian L, Bassalow R, Zaks D, Huntzinger C, Seuntjens J. IAEA-AAPM TRS-483-based reference dosimetry of the new RefleXion biology-guided radiotherapy (BgRT) machine. Med Phys. 2021; 48 (4):1884–1892.
43. Forster JC, Marcu LG, Bezak E. Approaches to combat hypoxia in cancer therapy and the potential for in silico models in their evaluation. Phys Med. 2019, 64: 145–156.

MIX
Papier aus verantwortungsvollen Quellen
Paper from responsible sources
FSC® C105338

If you have any concerns about our products,
you can contact us on
ProductSafety@springernature.com

In case Publisher is established outside the EU,
the EU authorized representative is:
Springer Nature Customer Service Center GmbH
Europaplatz 3, 69115 Heidelberg, Germany

Printed by Libri Plureos GmbH
in Hamburg, Germany